First Published 2016

Published by P&G Media
e-mail pds54@dsl.pipex.com

Design and page layout by Kriele Ltd
e-mail design_lamb@btinternet.com

Print/Production by 3-Point Ink, LLC
www.3pointink.com

BIG BUD

The Big Bud Tractor Story

Peter D. Simpson

The Big Bud Tractor Story is dedicated to all the great pioneers who have been involved with the development of the big 4WD articulating agricultural tractors.

Also to all those who worked on the development and building of Big Bud tractors from 1969 to present day.

Finally to all the men and women who continue to drive and maintain these iconic machines that have become legends in their own time.

Contents

Big Bud

The 'Big Bud Story' is the fourth book on big 4WD articulated tractors by Peter D. Simpson, who is a leading authority on big 4WD articulated tractors of the world. The 'Big Bud Story' is a specialist book, unlike his previous two books, 'Ultimate Tractor Power' volume one and two which are an A-Z of articulated tractors from around the world and the third entitled 'Big Iron in Pictures.'

Peter is a regular contributor to many monthly publications in England, Europe and N. America on all subjects related to farming, tractors, agricultural machinery and farm toys. To date Peter, in conjunction with a German film company and latterly Toy Farmer Magazine, LaMoure, ND, has produced nine authoritative DVDs featuring the big 4WD giants from around the globe. In the first of a new series on tractor pioneers, Peter D. Simpson, in conjunction with Toy Farmer, produced a detailed history of Wagner to Big Bud.

The 'Big Bud Story' charts the history of farming on the prairies and how the big 4WD articulated tractors came into being, detailing the origins of Big Bud tractors from the early beginnings in 1969 to present day.The history of these tractors is not as long nor complex as some other tractor makes, nonetheless Big Bud tractors hold a great interest and fascination as they are some of the world's largest agricultural tractors still at work today.

"I chose to write the story of Big Bud tractors for several reasons; mainly it fitted in with the series of books and DVD projects I am currently working on and secondly having an interest in big tractors world wide I could look at the subject matter from a different perspective and show the tractors in a true light.

"I have been involved with agriculture all my working life and I have seen some awesome sights around the world, but once you have been on board a Big Bud tractor or sat in the cab of the world's largest agricultural tractor, the Big Bud 16V 747, everything else pales into insignificance. Once the Big Bud bug bites you, you cannot shake it off, you always want more.

"For me the Big Bud tractor is the ultimate machine, if I miss my annual pilgrimage flying from England across the Atlantic and driving across the Northern States of America to see these giant machines and the new found friends I have made, life will never feel quite as good for the rest of the year."

Peter D. Simpson

I was raised on a farm 15 miles east of Havre, Montana. Through my early teenage years I developed an appreciation for some of the challenges farmers faced when making equipment decisions.

In 1960 my father and mother, Merle and Ernestine Harmon, sold the family farm and moved to Havre where they started a truck stop with fuel pumps, a service shop and a café. In 1961 Wilbur Hensler purchased land next to the truck stop and started selling Wagner tractors. By 1969 Wagner had sold its tractor production to John Deere leaving the Havre Wagner Dealership without a tractor to sell. That same year the Service Manager, Bud Nelson, and Wilbur Hensler started building their own tractor called "Big Bud". The name gave credit to Bud Nelson, the designer and production manager. These were very informative years for me, working next door to the Northern Manufacturing operation.

During 1975, at age 27, I purchased Northern Manufacturing from Wilbur Hensler and Bud Nelson. I changed the name first to Harmon's Northern Manufacturing, then later to Big Bud Tractors, Inc. The Big Bud Tractor production went from 12 tractors per year in 1974 to 80 tractors per year in 1979.

In 1977 we received the order for the 747 Big Bud tractor. It was a quantum leap from 450 horsepower, our largest production tractor, to a unit capable of 750 to 950 horsepower. Keith Richardson, who engineered most of the Big Bud models met the challenge. He engineered and oversaw the production of this tractor. Keith can be credited for the fact that there have been no major problems in the 10,000 hours and over 20 years of use of this tractor.

Robert and Randy Williams currently own this tractor. We are proud of the fact that people from all over the world visit the farm at Big Sandy, Montana to watch this unit work.

I want to thank all of the individuals who have supported and worked with us, including the current employees at Big Equipment Company, LLC and the past employees of Big Bud Tractor Company.

All of our crew would like to give special thanks for all the effort and excellent work undertaken by Peter D. Simpson in putting this book together.

Ronald M. Harmon

Ron Harmon.
Big Equipment Company, LLC

Horse Power to Tractor Power

To understand the requirement for high horsepower tractors on the farm that are capable of pulling 60 foot plus implements is complex, it is not just the past 50 years of agriculture that has to be understood but the previous 50 years also. The history and settlement of the great plains of North America more than 100 years ago set the scene for today's modern agricultural practices.

It is hard to understand, or comprehend that a pioneer settler moving into Montana in the mid to late 1800s could influence modern day agriculture. Understanding Montana's history and the vastness of one of America's Northwest States goes a long

CLOCKWISE FROM TOP:

> The Horse Shoe Bar beef herd on its way to the railroad in September 1896 in the Badlands of the Missouri River. Montana Historical Society, Helena.

> Taken from the Frank J Haynes collection dated 1879, this shows more than 16 teams of horses in a plowing gang at Grandin Farm in the Red River Valley. F J Haynes Photographic Collection, NDIRS-NDSU, Fargo.

way to understanding how the wheat and barley farms of this majestic state and the tractors that work the prairies have become legends of their time.

When the first pioneers moved north and west into Montana during the mid 1800s in search of gold, fur and a lucrative living, they headed into a wilderness occupied by Indian tribes, these tribesmen were content living off the land to feed themselves and their families.

The white miners and fur traders needed food, clothing and timber for housing. The settlers quickly followed establishing farmsteads, cultivating the rich fertile soil to grow wheat, barley and potatoes while many set up ranches raising cattle and sheep on the open rolling plains of this new found land.

From the beginning, the west has exerted a pull on the American spirit. The grassy interior between the Missouri and the Rocky Mountains was designated Indian Territory in the 1830s and was bypassed by emigrants on the Oregon Trail. By the 1850s huge land acquisitions had filled out the continental United States. The country's sheer vastness strengthened the conviction that the public domain rightfully belonged to the people.

Distributing land west of the Mississippi became an enormous project. The small settler farmer was unable to compete with the larger rancher farming operations, which led to the 1862 Homestead Act, an Act which declared that any citizen or intended citizen could claim 160 acres, which was one quarter square mile of surveyed government land.

Claimants had to improve their plot of land by building a home and growing crops. After five years, if the original claimant was still farming that quarter of land, it became his property. European immigrants followed, lured to America by railroad companies eager to sell off millions of acres of grant land and to provide farm-to-market transportation.

By 1900 more than 80 million acres had been handed out to 600,000 applicants yet quite often 160 acres or a quarter, as it became known, was not enough to feed and sustain a family especially in the drier northwest area of Montana.

Farming spread slowly across the state until the U.S. Congress passed the Enlarged Homestead Act of 1909. This legislation provided 320 acres of public domain land in Montana to settlers who pledged to cultivate half the land. Immediately after the passage of the Act, land offices processed between 1000 and 1500 homestead land claims per month.

Cattle and sheep ranches boomed continuing to take advantage of Montana's abundant grasslands. Over the following two to three years tens of thousands of homestead farmers moved into the state looking for inexpensive land.

The early farmers were quite content working their few acres of land with a team of horses, cultivating, seeding and harvesting. They grew enough to feed themselves, selling the surplus to the trappers and miners who passed through. Demand was quickly outstripping supply; the farmer was in great demand to produce more food.

CLOCKWISE FROM TOP:

> 15 teams of horses and men ready to start planting winter wheat.

> Five horses and one man made up a plowing team in the Red River Valley area of North Dakota.

Once the new generation of farmers acquired more land it became impossible to undertake all the work with one or two teams of horses as they had previously done. Neighbours would join together at busy times of the year combining teams of horses and implements when and where necessary to help meet the demand for increased food production. Early custom operators established themselves offering up to a dozen teams of horses pulling specialist equipment. Each team consisted on average of four horses capable of pulling a two-bottom plow. Early horsepower on this sort of scale was easy to work out,

CLOCKWISE FROM TOP:

> After cleaning out the cattle sheds, teams of horses are coupled to International Harvester No 4 Manure Spreaders head out to the fields.

> A labour intensive threshing team in the early 1900s.

10 teams of four horses equalled 40 horsepower. One team of horses could plow in the region of an acre a day.

The landscape was vast, each field covering up to 160 acres; to work the land even in those early days required considerable man and horsepower.

With the advent of the tractor, scenes similar to the earlier horse days were still evident. Again custom operators or contractors would move in with a fleet of mechanical horses. Fleets of tractors rated at 20hp plus now worked the land, a crew with 10 tractors could produce in the region of 200 drawbar horsepower, each tractor fitted with a two-bottom plow could plow in the region of one acre an hour.

The race for more horsepower was constantly on the tractor manufacturers' agenda with designers and engineers producing more powerful efficient engines with larger tractors coupled to larger implements required to work the vast rolling landscape to grow more crops with increased yields.

MONTANA HOME OF THE BIG BUD TRACTOR

Native Americans were the first inhabitants of the area to become known as the state of Montana. Tribes include the Crows in the south central region, the Cheyenne in the southeastern part of the state, the Blackfeet, Assiniboine and Gros Ventres in the central and north central area, the Kootenai and Salish in the western sector. The Pend d'Oreille were found around Flathead Lake, the Kalispel occupied the western mountains.

The Lewis and Clark Expedition of 1804-1806 was the first group of white explorers to cross Montana. Hard on the heels of the expedition arrived the fur trappers and traders. Trappers brought alcohol, disease and a new economic system to native populations.

The discovery of gold brought many prospectors into the area during the 1860s. The rapid influx of people led to boomtowns that grew rapidly then declined just as quickly when the gold ran out.

The State of Montana in the western United States of America is the northernmost of the Rocky Mountain States. Often referred to as the 'Treasure State' due to its mineral wealth, the name Montana came from the Spanish word meaning "mountainous" and was first used when the area was designated a territory in 1864. Montana became the 41st State on November 8, 1889.

Montana's land area consists of 93 million acres of rugged mountains, deep valleys and wide open plains. This is the fourth largest State in land area in the country behind Alaska, Texas and California and it is here where some of the world's largest production tractors were operated and built; names such as Wagner, Rite and Big Bud are synonymous with this State.

CLOCKWISE FROM TOP:

> Bindering on the Cooper Brothers' farm near Cooperstown, North Dakota where over 15 horse-drawn binders are harvesting wheat. Myrtle Portaville Photographic Collection, NDIRS-NDSU, Fargo.

> Working big acres meant using big tractors pulling big implements such as this scene with the 120hp Holt crawler pulling field cultivator, disc harrow, seeder and rollers on Campbell Farm, Big Horn County. Montana Historical Society, Helena.

Although many people consider Montana completely mountainous, two-thirds of the state is part of the Great Plains. From the majestic peaks of the Glacier National Park in the northwest to the comparatively level terrain near the eastern border. A large and sparsely populated State with an economy that has traditionally depended on natural resources and farming.

In 1853 the United States War Department made surveys for the construction of a railroad that would connect the Pacific Coast with the East with the population growth made possible by the completion of the railroads. Many of the state's newest citizens were lured by the silver and copper operations, coal and lead mining, and the development of the lumber industry. Montana attracted many immigrants during the late 1800s; its population grew from 39,159 in 1880 to 142,924 in 1890 and to 243,329 in 1900.

CLOCKWISE FROM TOP:

> Horses and tractors work together in Gallatin County near Boseman, the harvest scene was photographed by Albert Schlechten of Boseman, Montana. Montana Historical Society, Helena.

> World plowing record at Campbell Farm, Big Horn County, Montana. 640 acres was plowed in one day by 14 engines without one stop for mechanical trouble of any kind. Average gasoline consumption was three gallons per acre. 14, 30-60 Aultman-Taylor gas engines with an operator and an operator on each plow achieved this record. Montana Historical Society, Helena.

During the early 1870s early settlers in Montana included cattle ranchers who flourished in the western valleys as demand for beef in the new mining communities increased. Open-range cattle operations spread across the high plains, taking advantage of the free public-domain land. Some settlers decided that raising livestock on Montana's open ranges was more lucrative than prospecting for gold or growing crops.

Ranchers such as Nelson Story drove Texas longhorn cattle into Montana to graze on the rich prairie grasses before sending them to market south to Wyoming to sell them in Cheyenne.

Cattle and sheep ranches continued to take advantage of Montana's abundant grasslands. Passage of the Enlarged Homestead Act in 1909 brought tens of thousands of homestead farmers into the state looking for inexpensive land.

Ultimately the cattle ranchers were forced to give up the open range, which was fenced in with barbed wire, and the long Texas-to-Montana cattle drives came to an end.

The eastern two-thirds of Montana is part of the Great Plains, this region, known as the Missouri Plateau, is divided into two segments, the glaciated section in the north and the unglaciated section in the south.

Montana's rich soil produced some of the nation's best wheat and from 1909 to 1916, Montana was blessed with above-average rainfall which saw its wheat production rise from 11 million bushels in 1909 to more than 42 million in 1915. In addition, wheat prices increased during World War I (1914-1918), which increased demand for wheat, and Montana farmers prospered.

In 1917, however, this prosperity was suddenly reversed by a ruinous drought. The farmers compounded the effects of the drought by planting crops on lands better suited to grazing. The soil, dry from lack of rain, turned to dust and blew away;

CLOCKWISE FROM TOP:

> Farming the large prairies was a lonely and a slow job with one tractor and harvester. Al Lucke Collection, Montana State University, Havre, Montana.

> In 1945 nine tractors each with one-way disc plows covering approximately 14 foot each work in unison. Montana Historical Society, Helena.

in addition, hordes of grasshoppers appeared and ravaged the remaining crops. To make matters worse, international wheat prices fell after World War I.

In the early 1920s land values dropped, half of the farm mortgages were foreclosed, and 20 percent of farms were abandoned as families sought better opportunities elsewhere. The homestead "bust" forced many farmers to abandon Montana.

Montana's post-World War I depression extended through the 1920s right into the Great Depression of the 1930s. Montana's farmers learned by experience the need for conservation measures. The droughts of the 1930s brought economic ruin to many ranchers and farmers, who for years had overstocked the ranges and plowed grasslands suitable only for grazing. Since then, strip farming, contour plowing, and other soil-conservation practices have been widely adopted.

Post World War 2 Montana had been characterized by a slow shift from an economy that relied on the extraction of natural resources to one that was service-based. Agriculture, while dependent on weather, a declining workforce and international markets, had remained Montana's primary industry throughout the era.

The total number of farms and ranches has decreased in recent decades, but the size of the average unit has steadily increased. The average size of Montana's 23,100 farms and ranches in the mid-1990s was 2,584 acres. Altogether there were 59.7 million acres of farmland, one quarter of the

CLOCKWISE FROM TOP:

> Botler Creek, Oregon on Jasper Merger's ranch, this Wagner TR-14 pulling a 105 foot rotary plow in 1960 was operated by one man. The Wagner Brothers, Oregon.

> Such was the versatility of the big Wagner WA-14 tractors and the strength and longevity of Versatile's seeder boxes, this unit was still working in 2005 where one man could seed in the region of 30 acres an hour.

farmland was used for crops, while almost all the rest was rangeland or pasture.

Wheat, the leading field crop, provides about two-fifths of the state's farm income and about three-quarters of its income from crops. Most of the high quality wheat is grown in the plains section, with winter wheat being grown mainly in the area north of Great Falls in an area known as the 'Triangle Area' or the 'Golden Triangle'.

Montana ranks third among the states in wheat production growing 154 million bushels in 1999, behind only Kansas and North Dakota.

In the mid-1990s livestock, largely beef cattle, and livestock products accounted for slightly more than one-half of farm income.

During the late 1990s and into the early years of 2000 the northeastern area of Montana saw rainfall drop considerably with many wheat crops continually failing. More and more farmers headed toward direct seeding and 'chem fallow' reducing the need for traditional cultivation practices.

More farm workers and farmers left the land and headed into the larger towns and cities to find a more lucrative living. The farmers that remained had to farm larger acreages with lower labour input with reduced yields and a bleak future. Those who operated high horsepower tractors that used to work in the region of 400 hours per year saw working hours reduce to less than 100 hours per year with lo-till and direct seeding systems.

During the 1980s and 90s two or three tractors would have been used at seeding time, one cultivating with the second seeding, the third machine would have cultivated the summer stubble to kill germinating grain and weed seeds. The use of Roundup and other 'chem fallow' operations has eliminated the need for summer fallow cultivation. Direct seeding or no till has reduced the need for the second tractor cultivating the land in front of the seeder.

Reduced yields coupled with lower wheat prices forced the farming community to review its overall plan in early 2000. The requirement for one large 4WD tractor to undertake the full farming operation had become a necessity with many big tractors parked up to be used as a back up power horse or sold. The Great Plains farm economy has undergone such a huge change that traditional farming in Montana will never be the same again.

CLOCKWISE FROM TOP:

> Two Big Bud 400/30 tractors fitted with Flexi-Coil cultivators can work with ease several hundred acres in a few hours.

> The Big Bud 747 with its 86 foot field cultivator can work in the region of one acre every minute.

CHAPTER 2

The metal tracked tractor

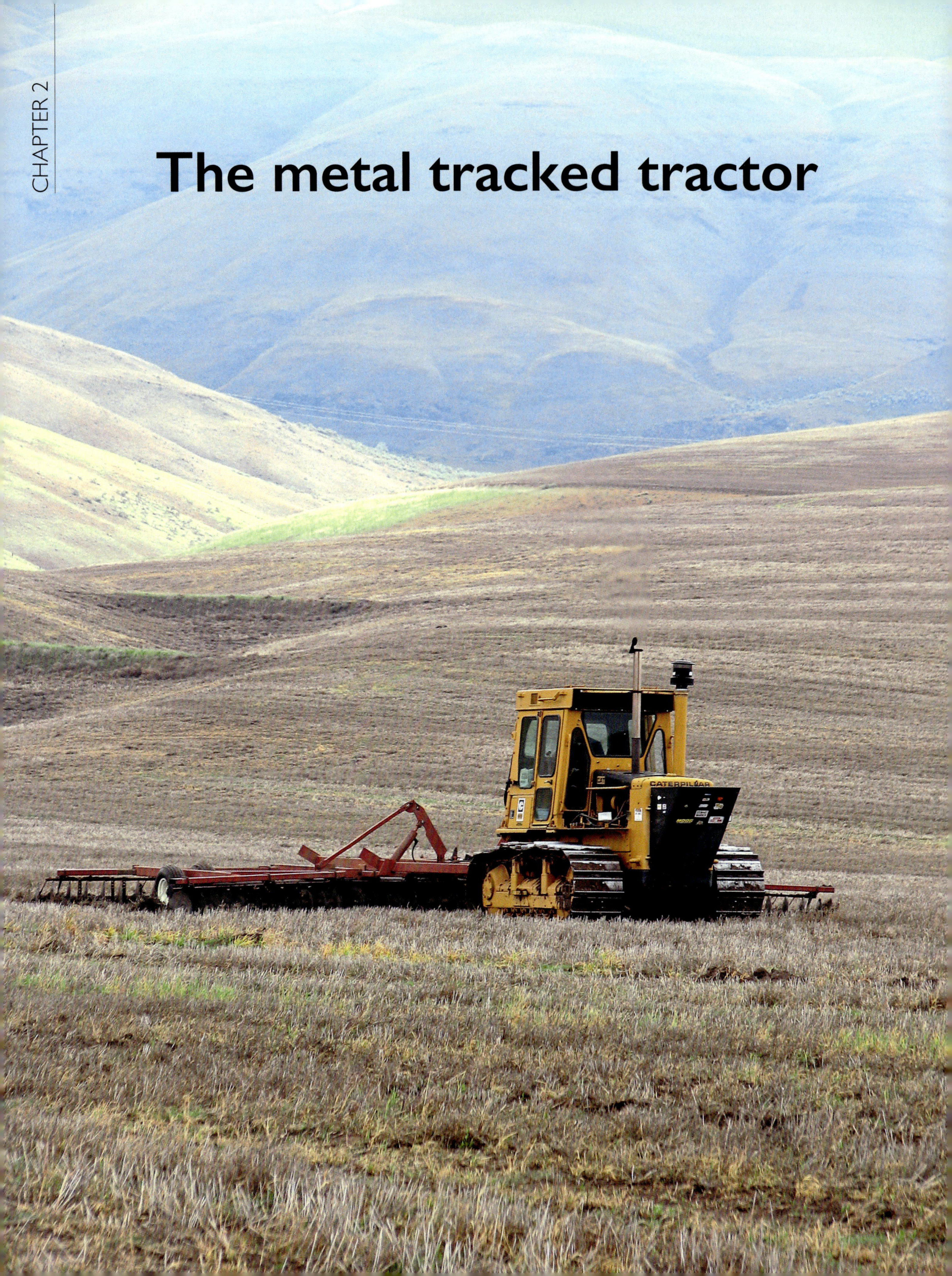

The rolling hillsides of California, Oregon, Washington and Montana were often too steep for conventional 2WD tractors to traverse making cultivation for crops difficult. These slopes remained in pasture for cattle and sheep to graze while the flatter plains were being opened up for grain production on a large scale.

With the introduction of more powerful 2WD tractors during the 1940s, the metal tracked crawler tractors also became more powerful proving their worth on much of the untouched fertile foothills of the Rocky Mountain Range and other hilly ranges. With the dominance of the crawler tractor grain production quickly spread across the rolling foothills of these mountainous regions.

The big high horsepower wheeled tractors could not compete on rough hilly terrain, across peat, wet or soft ground; the crawler tractor became a favourite. The tracked tractor was a reliable workhorse during the 1940s and 50s on the rolling great north west wheat belts of the United States. Quite often this type of machine was the only type of tractor that could safely work on the steep hillsides.

For many years tracked tractors remained supreme on this type of terrain; a low centre of gravity coupled with a wide footprint gave maximum traction, a considerable degree of flotation, and a low ground pressure were just a few of the advantages of these machines.

CLOCKWISE FROM TOP:

> The tracked tractor enabled farmers to traverse and cultivate land that previously had only been used for grazing. The 61.56hp International Harvester TD-14 and McCormick 51 Harvester Thresher can be seen operating on a 60% gradient with ease.

> An International Harvester TD-9 tracked tractor cultivating hilly ground in the foothills of the Rocky Mountains.

> On some very steep hillsides, especially the Palouse in Washington State, tracked tractors still dominate and are the only machines that can work safely to this day.

With the introduction of the Wagner 4WD articulating tractors in the mid 1950s the farmer was offered an alternative, the tracked tractor's dominance on difficult terrain was challenged for the first time.

It took a while for farmers to accept this new type of wheeled tractor, but once tested it was only a matter of time before tracked tractors virtually disappeared from the hills.

The 4WD articulating tractors remained supreme on this difficult terrain unchallenged for nearly half a century when the new breed of tractors proved their worth during the 1990s and early 2000s.

CLOCKWISE FROM TOP:

> Caterpillar D6 and D8s rated at 70 drawbar hp and above could work land that tractors of twice the power and size could not work.
> During the late 1930s tracked tractors such as the Allis-Chalmers Model M, rated at 22hp at the drawbar, were invaluable in opening up pasture land for crop production.
> Caterpillar tracked tractors had been and still are today popular machines in agricultural land reclamation and crop production.

CHAPTER 3

Tandem tractors

With the growth and expansion of farms the race for continued power with improved traction went on year by year. Each year the leading tractor manufacturers provided the optimum power to weight ratio introducing the ultimate machine only to be beaten the following year by more power and improved design from a competing manufacturer.

As crawler tractors began to dominate the rolling plains working where many smaller wheeled tractors could not go, another form of increased power and traction began to emerge. Joining two tractors together in tandem began to generate considerable interest during the 1940s, this was the evolution of the agricultural 4WD articulating tractor.

Joining two tractors together formed a crude but effective 4WD system that led to the first true 4WD tractor from the Wagner Brothers.

This concept started on the farm with a farmer tinkering and designing a simple tractor hitch so that he could couple two tractors together with the end result being effectively a 4WD tractor that articulated. With the use of two equal sized engines, pulling power of the two tractors joined together could be more than doubled.

When not in use the tandem tractor could be separated back to two smaller tractors, each ready for work in its normal power range and duties. Quite often a single big tractor can be wasteful on light duties.

During the 1950s through to the 1970s many farmers made a living by hooking up two and sometimes three tractors. By the mid 1950s tractor manufacturers offered kits and dealer assistance to hitch two tractors together. This cheap alternative form of power meant that tractors could travel where a conventional 2WD had previously found it difficult. The front unit could pull the rear unit through soft going while the rear

ABOVE:

> Charlie Inman from Havre, Montana owned a John Deere 830 and a JD 820 which he coupled together to easily produce 150 drawbar hp.

unit could push the front unit.

With increased power came the need for larger implements; smaller implements were also coupled together with one operator often more than doubling previous work rates.

There were three basic ways of hooking tractors together in tandem.

The front unit was a complete tractor with a steerable front axle, the rear unit had its front axle removed, the two were coupled together by an A frame. The operator usually worked from the front unit. A three-axle, six-wheel machine.

Two tractors with both front axles removed were coupled together by means of a turntable fixed under the front end of the rear tractor, steering was by hydraulic rams. The operator usually worked from the rear unit. A two-axle, four-wheel machine.

An inexpensive way was to leave both front axles on the tractors, the rear unit's front axle sat in a frame fitted to the rear of the front tractor. The operator worked from the front unit. A four-axle machine with eight wheels yet only six wheels had actual ground contact.

CLOCKWISE FROM TOP:

- > The Dufners from Iowa built and used many tandem hook ups. Here the family's first tandem in 1962, a John Deere 720 at the front and a JD Model D at the rear, together they produced around 110 drawbar hp capable of pulling an eight bottom plow.
- > Plowing in 1972 with a John Deere 820 at the front and a Model D at the back producing around 135 drawbar hp pulling two five bottom plows hooked together.
- > Hooking two similar sized tractors together in tandem gave twice the available power, it was inevitable that two or three implements had to be hooked together to utilise the increase in power, almost doubling previous work rates.

The most cost effective and most productive way of joining the two tractors together was by removing the front axle of the rear unit, which also gave good ground clearance. This way offered very few major engineering problems compared to both front axle removal.

The most modern tractor was usually used on the front; the second or rear tractor was usually a little used machine of older years in good mechanical condition. This method of joining two tractors together caused very few steering problems as the operator steered the front tractor conventionally with the rear unit following behind, similar to as a trailer would.

Perhaps the most difficult part of the operation was the controls, both tractors had to be operated from the front unit; these were often designed to suit the individual skilful operator.

A further use for hooking two tractors together was when power was required to operate a piece of trailed or mounted

CLOCKWISE FROM TOP:

> Hooking two tractors together became common practice in the 1950s and 1960s often doubling horsepower and reducing labour on the farm.

> When Joe Wilson from Illinois joined his two Allis-Chalmers together he modified the tinwork. Joe removed both sets of front wheels using a pivot configuration with steering rams more like a true articulating 4WD tractor.

> A simple way of hooking two tractors together was by a drawbar method, both tractors were operated from the front seat and the rear tractor simply followed the front tractor.

equipment. The front unit solely provided power for forward movement whilst the rear unit provided power to operate the equipment. If the driving unit's power decreased or the tractor wheels started to spin, the power to the implement remained constant.

The tandem unit was a good, cheap form of 4WD horsepower; it had its advantages and disadvantages. When not in seasonal use a unit could be split and put back to two conventional tractors that could be used for yard work and other light duties.The disadvantage was that the tandem unit was difficult to operate, both units had to work as a team as two horses would have had to do, the tandem unit in the correct hands was invaluable, in the wrong hands two tractors were wasted.

CLOCKWISE FROM TOP:

> The tandem tractor utilizing two John Deere tractors was approximately 150hp and could pull a 34 foot cultivator at around 5½ mph covering 22 acres per hour using in the region of one gallon of fuel to three acres. Here, two three furrow John Deere plows are hooked together and pulled with ease.

> The front axle of the JD 820 was removed and an A frame was constructed.The drawbar on the JD 830 was strengthened, a thick pin was constructed, when the rear tractor was coupled to the front tractor all was secured by a safety chain. Keeping the front axle on the front unit required less engineering expertise for the steering system as the whole unit was steered by the front axle. An early form of a 4WD articulating tractor was engineered.

CHAPTER 4

WAGNER TRACTOR

AGRICULTURAL MODELS

MODEL TR-14 SHOWN HERE

The BIG HINGE Makes the Big Difference!

Wheels Always Track!

No Loss of Power in Turns!

4-WHEEL DRIVE
PLANETARY DRIVE

4-WHEEL POWER STEERING
BOTH AXLES OSCILLATE

CENTER PIN STEERING - HYDRAULIC WITH FLOW DIVIDER

POWERED BY FAMOUS CUMMINS DIESEL ENGINES

OPTIONAL ITEMS INCLUDE CAB, AIR CONDITIONING UNIT, HEATER AND DEFROSTER — FOR GREATER OPERATOR COMFORTS

WAGNER TRACTOR, INC.

8027 N. E. KILLINGSWORTH P. O. BOX 7444 PORTLAND 20, OREGON

Wagner Tractor Inc

Establishment of the four-wheel drive articulating tractor

OWNERS AND DIRECTORS

WAGNER TRACTOR INC.

SEVEN WAGNER BROTHERS

AND J. BURKE LONG

During 1922 Eddie Wagner and his six brothers, Bill, Gus, Walter, Harold, Irvin and Elmer from Portland, Oregon started a company called Mixermobile. Between them they designed a mobile machine that could mix and place concrete on site, hence the company name Mixermobile. From this small beginning the brothers became pioneers in several major areas, which included forestry machinery, wheeled loading shovels, mining equipment and other four-wheel drive machines associated with industrial and agricultural uses.

Wagner Tractor Inc, was incorporated in Portland, Oregon in February 1954. Its owners consisted of the seven Wagner brothers of Portland and J Burke Long, previously of California, who became a resident of Oregon. The active officers consisted of Irvin Wagner, Elmer Wagner and Walter Wagner along with J Burke Long.

At that time Wagner Tractor Inc. products consisted of three types of tractor; the Industrial Models for road building, earth moving and general construction work, the Logging Models for the forestry and logging industry and the Farm Models for agriculture, thus reaching the market of three very large but distinctive industries.

The Wagner tractors were not a 'backyard development' without substantial background, but were backed by the reputation and know how of men who had spent the previous 20 years in the design and production of new and outstanding heavy-duty equipment. Their previous developments included such well-known machines as the Mixermobile, Towermobile, Scoopmobile, Due-Way Scoops and Lifts, Buggymobile, Telescopic Lifts, Foldaway Lifts, Dozermobile and Stationary Mixers.

Of these previous developments the Dozermobile, initially manufactured in 1950, was the first machine to be built by employing the principle of dual unit construction, which became the basic principle of the Wagner tractor. This was made possible by joining two units, commonly referred to at that time as bogies, by the use of the Wagner designed and patented Centre Hinge Assembly, later to be known as the Pow-R-Flex Coupling.

This resulted in free axle oscillation of both axles and also

CLOCKWISE FROM TOP:

> A period photo from the 1960s.

> From left to right Bill, Guy, Eddie, Walt, Harold, Irvine and Elmer with J Burke Long above.

Another WAGNER RECORD!

Wagner Owner Saves $3,842.10 In One Year Farming 2000 Acres!

One Wagner TR-9 Tractor Does Work of Three 2-Wheel Drive Tractors and Saves $3,842.10 Annually

ROLAND WRIGHT LIVES IN MOORE, MONTANA. WITH HIS FATHER BURTON WRIGHT, THEY FARM 2000 ACRES IN CENTRAL MONTANA, USING ONE WAGNER TRACTOR. PREVIOUSLY SAME WORK REQUIRED THREE TWO-WHEEL DRIVE TRACTORS. AT RIGHT: ROLAND AND HIS WAGNER TR-9.

Roland Wright says, "In covering 14,643 acres with my Wagner in repeat operations consisting of Spring land preparation, Spring seeding, Summer fallow and seeding Winter wheat, my labor and fuel cost was *only 16c per acre,* but doing same work previously with three 4-5 plow, 2-wheel drive tractors my cost was 42½c per acre. My Wagner averaged 16.14 acres per hour; uses only 0.29 gallons fuel per acre. No wheat farmer should be without one."

COMPARISON OF WRIGHT'S WAGNER TR-9 WITH 3 TWO-WHEEL DRIVE TRACTORS

	COST SUMMARY TR-9	COST SUMMARY THREE 2-WHEEL DRIVE TRACTORS
TOTAL ACRES WORKED	14,643 Acres	14,643 Acres
TOTAL FIELD HOURS	1,080 (1 Tractor)	2,928 Hours (3 Tractors)
TOTAL FIELD DAYS — 10 Hours Each	108 Days	292.8 Days (3 Tractors)
TOTAL FUEL USED, No. 2 Diesel	4,303 Gals.	10,248 Gals. (3 Tractors)
LABOR $15 DAY PER OPERATOR	$1,620.00	$4,392.00 (3 Operators)
FUEL COST @ 18c GALLON	$774.54	$1,844.64 (3 Tractors)
TOTAL LABOR AND FUEL	$2,394.54	$6,236.64 (3 Tractors)
LABOR AND FUEL PER ACRE	16c per acre	42½c per acre

WAGNER SAVINGS PER ACRE — 26½c
WAGNER SAVINGS PER YEAR — $3,842.10

For additional information regarding Wagner Tractors, write to:

WAGNER TRACTOR, INC. BOX 7444, PORTLAND 20, OREGON

Front and rear units are free to articulate 40 degrees in either direction for tight turns.

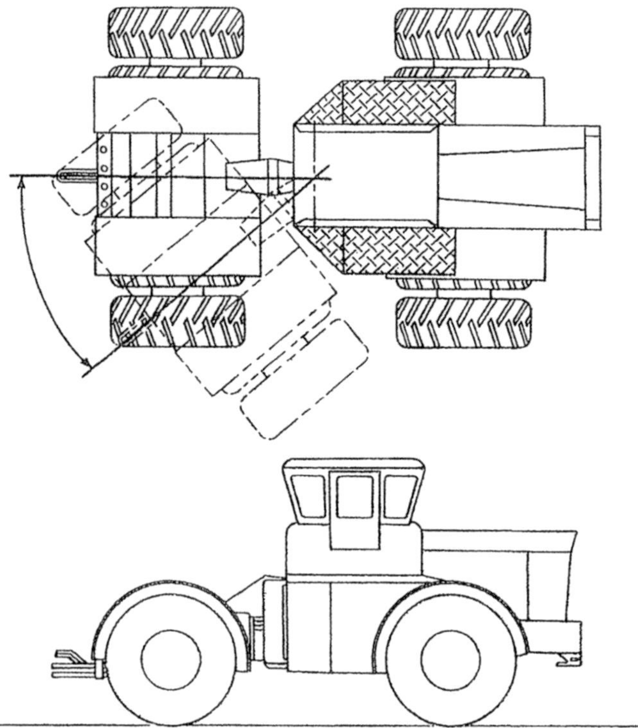

A 60-to-40-percent weight ratio on front and rear units levels out to nearly 50-50 under load.

Both tractors hug sidehills and operate with excellent stability on a 30-degree tilt.

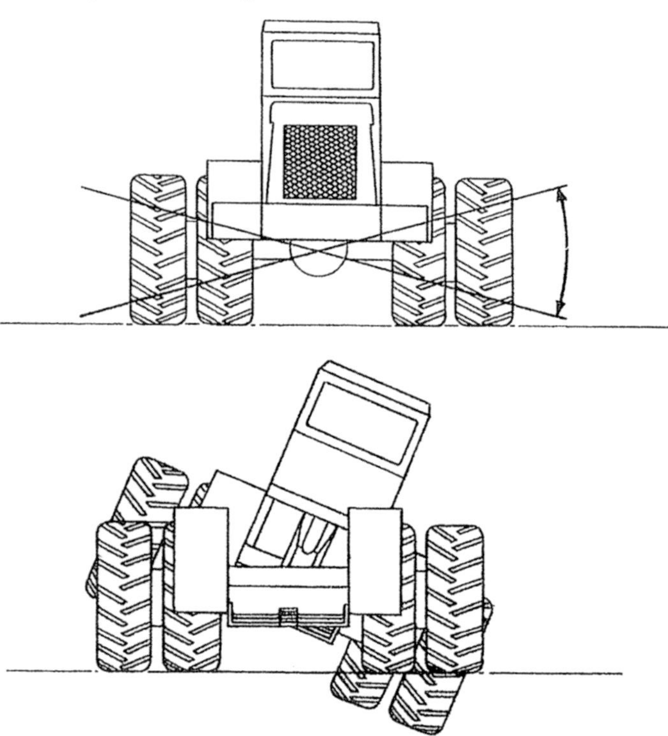

Units can oscillate 15 degrees in either direction to keep all wheels on the ground.

provided true 4WD and four-wheel steering with resulting advantages never before seen on any other make of tractor.

Previous experimentation had taken place by the Wagners as early as 1947 with the application of 4WD in an agricultural tractor built on a single unit chassis and semi-rigid frame. Its limitations were immediately apparent in trying to travel over rough terrain, loose footing or mud where four-wheel steering could not be successfully applied. In reality, this experimental model was only a true 4WD on level ground where it was least required.

The answer to the problem in tractors at that time was apparent from the success of the earlier Dozermobile dual-unit construction machine and in 1954 Wagner was determined to enter the very lucrative agricultural field of manufacturing.

The problem to solve was to keep equal weight distribution on all four wheels at all times regardless of terrain or footing and to provide 4WD and four-wheel steering that would permit wheels to track and continue to furnish full power and traction at any degree of turning. No other wheel type tractor had ever accomplished this and, of course, it was possible in crawler type tractors which turned by use of steering clutches yet lost much power when turning.

The Wagner answer was the relatively simple but patented system of using the Pow-R-Flex Coupling, which allowed free oscillation of both axles within any pre-determined limits by the unique horizontal oscillation tube with the drive shaft through the centre.

It was soon discovered in actual operations that the purpose of allowing the tractor to travel amazingly rough terrain became secondary to the tremendous traction advantage on loose footing; the effect being that any wheel which started to spin was absolutely free to dig in and carry its normal share of weight rather than shifting extra weight to the wheel that was theoretically in trouble; thus allowing it to bear down and rapidly gain a tremendous tire to ground contact advantage that instantly restored its tractive efficiency to the equal of the wheels on firm footing.

It further transpired that the Wagner dual unit construction allowed tractors for the first time to travel rough hilly terrain beyond all expectations without tipping over, as would any single unit tractor. This was due to the fact that when one wheel dropped beyond the normal tipping point the dual bogie stabilized it upon contact with the oscillation stops.

The Wagner patented Pow-R-Flex Coupling also provided the ultimate of simplicity in articulated steering (hydrostatic powered) in that only two pins and one hydraulic cylinder accomplish steering of all four wheels. These take the steering action equally through two lightly loaded drive shaft joints

WAGNER HINGED HOOD

BOTH AXLES OSCILLATE

POW-R-FLEX COUPLER

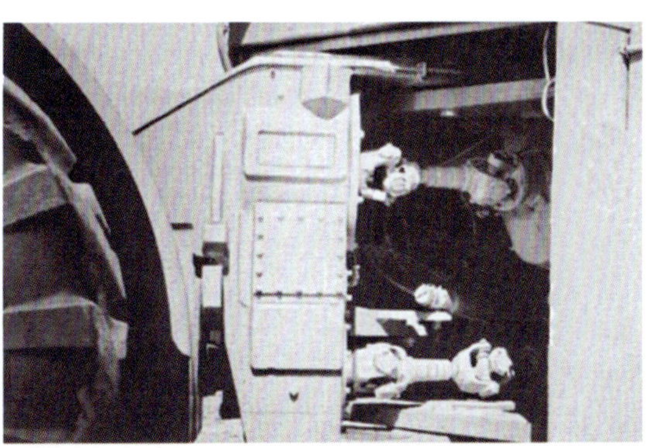

TWIST & TURN — Wheels always track

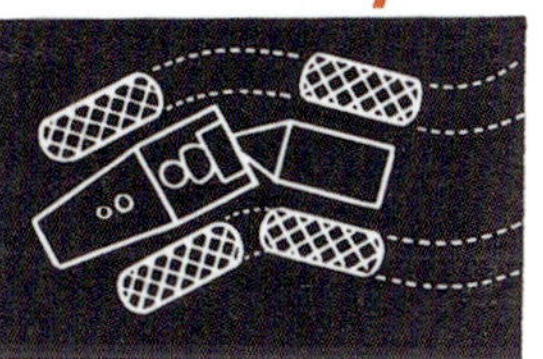

— Every wheel pulling

rather than the conventional method of steering knuckles in the final drive wheels. The obvious advantages were a short turning radius with full power of all wheels in any degree of turning with all wheels tracking and at the same time eliminating steering clutches without loss of power at any time.

Wagner tractors for construction were first manufactured and used in 1951. As with all new products certain changes and modifications became apparent from the initial tests of these early units and by late 1954 the Wagner agricultural tractor had passed through all stages of testing and any changes needed for this industry were incorporated to reach the stage of virtual perfection.

Many of the Wagner features such as the 4WD, four-wheel steering, two-axle oscillation and planetary drive had been proven in other products from the Wagner brothers, of which thousands had been manufactured.

The Wagner tractors were filling the need for a tractor with higher speed, greater mobility and less maintenance time and costs. These advantages were quickly proven in the field plus their ability to continually out perform the crawler type tractors under nearly all conditions.

VISTA DEL LANO FARMS

Outsiders were invited to the Vista Del Lano Farms in Cantua Creek, California, in which no Wagner tractor personnel participated, where they made an independent and impartial test and comparison in August 1955. A Wagner TR-14 tractor was started on an 80 acre field with two six-bottom 28 inch disc plows. A comparable sized and priced crawler tractor was placed on an adjoining 80 acres with identical implements and settings. The Wagner tractor plowed and completed the 80 acres while the crawler tractor plowed less than 50 acres, with the Wagner using five gallons less fuel in the same time.

The mobility of tyre-mounted tractors over crawlers was a well known and established fact. In addition to higher speeds, the cost of operating and maintaining was considerably lower. Job to job moves could be made across pavements, over kerbs, sidewalks and railroad tracks and on moves saved time, bother and expense of locating a trailer, moving in extra manpower and transportation equipment, loading, blocking, and unloading. It also permitted the tractor to return to the farm sheds and workshops quickly and easily for storage or repair and maintenance instead of remaining at job sites as a crawler would have done.

In 1954 some difficulties appeared with the new tractors that stemmed from a few basic problems, all of which were completely remedied. Diesel engine failures were experienced when using two well known makes of diesel engines; this was

quickly overcome by exclusively adopting Cummins engines.

Use of a standard truck straight drive axle for application of 4WD bought about some failures in the front axle assembly due to the fact that it was inverted. This was corrected by the use of Clark industrial paired axles, which contained properly cut reversed gears in the front differentials.

Lack of planetary drive axles resulted in torque to the drive train, which caused breakage due to the high-powered engines. This was eliminated by adopting as standard in all tractors, planetary drive axles with a 3 to 1 or more reduction in the final drive thus reducing by a minimum of 65% all torque to the axles, differentials and drive line assemblies.

Original design of the transfer case resulted in some failures to the reversing gears then contained in the gearbox and some drive chains broke due to the increased diameter of the driving sprockets; by increasing sprocket diameters and increasing the chain capacity this overcame the problem. Eliminating entirely the reverse gears from this case, reversing the chain tensioner to give an outward rather than an inward pressure corrected this.

The original centre hinge casting was too narrow in the Wagner-14 Series models but this was easily corrected by increasing the bearing spread and casting the entire assembly much heavier.

The size of the transmission was highly underestimated but was corrected by changing to a heavier transmission. Other improvements included heavier ply tires and a substantial increase in engine displacement to give added horsepower at lower rpm.

These early developments and improvements were made after much testing, study, considerable research and engineering which was contributed by some of the greatest corporations in America whose time proven components were used in Wagner tractors.

During a relatively short period of two and a half years, 4WD articulating Wagner tractors had gained an outstanding reputation and public acceptance in the northwest section of the United States and Western Canada. This was attested to by the fact that during this period 43 dealers in this area had purchased in excess of 350 Wagner tractors with a sales value exceeding six million dollars. Many owners of Wagner tractors went onto purchase a second and a third model.

The initial introduction of a new unit of any kind in any area is always the most difficult, but the Wagner Brothers had a

CLOCKWISE FROM TOP:

> Once the high horse powered Wagners appeared, the building of equipment to be towed behind also quickly grew. Here a 1960 TRS-14A is pulling a field cultivator just under 60 foot wide.

> In 1960 this TR-14 pulling an 84 foot box seeder was caught on camera.

Wagner Owner Performs 3 Operations at once with Big $avings! Doubles production of Replaced Crawler!

One Wagner TRS-14 Tractor pulls 36-ft. Graham-Hoeme, plus 36-ft. Rod-Weeder attachment, followed by 36-ft. Hoe Drills at one time, covering 175 acres in 10 hours. Also doubles production of replaced crawler tractor pulling 27½-ft. Off-set Disc. Eliminates one hired man.

MANLEY KIRKEBY
Shelby, Montana

RIGHT: Wagner TRS-14 pulling 36-ft. Graham-Hoeme, 36-ft. Rod-Weeder attachment, and three 12-ft. Drills in one operation.

LEFT: Wagner TRS-14 pulling five 8½-ft. Nobel Blades, Model 'M' covering 36-ft. to 40-ft. swath.

BELOW: Wagner TRS-14 pulling 27½-ft. off-set Disc.

Says:

"I bought my Wagner in the Spring of 1956 and at that time traded in my TD-14 crawler tractor.

"My Wagner pulls nicely at 4 M.P.H. in 5th gear a 36-foot Graham-Hoeme, followed by a 36-foot Rod-Weeder attachment, and this followed by 36 feet of Hoe Drills consisting of three 12-foot drills. This covers 175 acres in 10 hours. With my crawler I could only cover 90 acres in the same time.

"My Wagner pulling five heavy duty 8½-foot Nobel Blades, Model M, covers on the dry side 200 acres in 10 hours rolling time, but my crawler could only pull three of these same blades and did only 100 acres in the same time. On this type work my Wagner has doubled my production as against the crawler.

"The Wagner pulling a 27½-foot off-set disc at 5 M.P.H. covers 160 acres in 10 hours but the crawler could only pull 20-foot to 24-foot maximum discs and covered only 80 acres in the same time.

"My Wagner eliminated the vibration of the crawler tractor and speeded up the work by getting it done at the time it was needed. Now I do all my own work without the need of hiring one other man as before, and I have eliminated all night work."

For additional information regarding Wagner Tractors, write to:

WAGNER TRACTOR, INC. BOX 7444, PORTLAND 20, OREGON

compensating advantage sales wise in that once having created the desire for the 4WD Wagner tractor, this was not too hard to accomplish due to the outstanding performance and mobility. In other words, once they sold the farmer the idea of 4WD rubber tired tractors he had no competitive manufacturer or dealer to go to, to shop around or get a better deal, as often happened with competitive equipment when many companies produced basically identical tractors varying only in name. These tractors were unique and no other manufacturer could offer anything nearly as effective or versatile as a Wagner tractor.

The Wagner tractor was built of the highest quality components and materials money could buy. Consequently the unit price was slightly higher in some instances than the nearest competitive tractors of equal drawbar pull but compared as it should be on work produced per hour per dollar, it was much lower priced than any other tractor.

The Wagner rubber tired tractor could not, nor was it intended to, do everything a crawler could do; but everything a Wagner could do - it did at 20 to 50% less cost, it achieved 90 to 100% of all farming operations done with crawlers.

Elimination of the track friction power loss made the terrific fuel savings possible, plus the very efficient diesel engines manufactured by the Cummins Engine Company which were used exclusively in Wagner tractors. On older heavy crawlers it took much of the horsepower just to move them with no load in the high gears.

By achieving more work per day with less manpower, at a lower cost of operating and maintaining, which resulted in a vastly greater return per dollar investment, the Wagner was able to accomplish the same amount of work or acreage in a much lesser time than a crawler.

The Wagner Brothers did not claim to pull a larger load than a crawler, rather that they would out produce them with tremendous savings by properly matching load to traction.

The only rule to follow to learn what a Wagner could do and what it would save was to always match the implements well within the tractive limitations of rubber tires and always keep

CLOCKWISE FROM TOP:

> Long before laser land levelling became popular for flood irrigation, levelling fields was left to the judgement of the operator. Early Wagners such as this TR-9 were supplied with or without cabs.

> With high horsepower and big equipment such as this TR-14 with 48 foot offset disc plow, work rates increased; one tractor and one man could work fields that had previously required three tractors and three operators.

in reserve traction for the occasional soft spot, sand or hills.

Whereas the crawler had always operated with maximum loads and slow speeds to prolong the track life, the opposite principle was used with the Wagner tractor to get maximum production, by faster speeds with correctly matched implements. A tire had tremendous traction until it started to spin, so the only way to deprive oneself of the low cost per acre high production performance with a Wagner was to load beyond tractive ability.

Maximum horsepower ratings considered alone were without meaning. Ratings were studied at various speed ranges to measure their effectiveness in 'pull' effort at the most favourable operating speeds. Power at maximum drawbar was seldom used for tractive effort, for farming probably did not exceed 5% of all work.

The most effective cost and power saving combination permitted working at the highest practical speeds which job conditions permitted, and that was the principle on which the Wagner was designed and intended to work. There were very few times on farm operations where crawler maximum drawbar at low speeds gave a production advantage, compare those few times with a large percentage of times where extra speed paid off in profit saving.

The question therefore 'What will it pull?' must be replaced with 'How many more acres per day will it cover with implements it will pull and at what savings?'

When the Wagner brothers, Walter, Elmer, Irvin and J. Burke Long started developing agricultural articulated tractors in 1953/54, they were described as pioneers, visionaries and innovators and, without a doubt, true craftsmen. No one could ever fault the quality of their own construction. Good heavy steel, square cut with beautiful welding all done by hand.

The first full production Wagner tractors were seen in late 1955 early 1956 with the introduction of the TR6, TR9 and the TR14. The TR6 was fitted with a Waukesha model 190DLB six-cylinder engine rated at 64hp at 2200 rpm, its power approximately matching an IHC Super WD9 crawler.

The very first TR9 was fitted with a Buda engine, model

CLOCKWISE FROM TOP:

> Moving big tractors requires powerful transport.

> Dave Curtis of Rite tractor fame stood in front of what is believed to be one of only two Raygo Wagner W-24 tractors built.

> An FWD Wagner WA-4 pulling an eight-furrow plow makes light work of the job in hand.

6DT 468 rated at 85hp at 1600 rpm. The TR9 was marketed as being equal to an IHC TD14 crawler. The TR6 and TR9 were of similar appearance. The larger TR14 was fitted initially with a Buda engine and latterly a Cummins engine rated at approximately 160hp. The TR14 was reported to be tough competition for the Caterpillar D7 and D8 tracked tractors.

CHANGE OF OWNERSHIP

Part of the Wagner Corporation changed ownership in 1961 when the tractor division was purchased by the big eastern fire engine, truck and axle manufacturing company the FWD Corporation in Wisconsin (FWD stands for Four Wheel Drive). The new tractor company became known as FWD Wagner Inc., with the logging machinery and agricultural tractor manufacturing operations moving into bigger premises in Portland.

In 1961 the FWD Wagner WA tractors were introduced, the WA-4, WA-6, WA-9, WA-14, WA-17 and the WA-24 with the 6, 9, 14 and 24 models matching the earlier TR models. Several of the first WA-4 tractors were fitted with a three-cylinder Detroit Diesel 3-71 engine rated at 98hp. These tractors were short and stubby, quick and nimble. A three-point hitch was available on the WA-4. In 1962 the revised WA-4 was fitted with a four-cylinder GM 4-53 engine rated at 120hp at 2500 rpm. As the WA-6 was so closely matched to the new WA-4, production of the WA-6 ceased.

FWD was not an agriculturally orientated company; it was more heavily involved in industrial machinery. During the mid-1960s FWD sold all of the tractor manufacturing business to the Raygo Corporation of Minnesota. Raygo at that time was manufacturing a self-propelled drum type vibratory compactor and the company was looking at ways of expanding its operation. The wheel type compactor was a revolutionary machine to the road building and garbage sector and was developed by FWD Wagner.

Several of the FWD Wagner WA-24 tractors became known as Raygo Wagners. The Raygo Corporation became big in industrial articulated tractors, compactors and wheeled loaders which enabled the company to expand into further areas until the demise of Wagner agricultural tractors with the John Deere deal in 1968.

When FWD purchased the Wagner Tractor line the Wagner compactor had just completed a year of testing and favourable reports came in from owners, State and Federal Agencies and testing labs; Wagner tractors had produced previously unheard of compaction results.

At about the same time Raygo had produced the vibratory drum type compactor with equal success in its class. FWD decided to concentrate its efforts to the fire truck line and its already established axle and truck manufacturing lines. Raygo was well aware of the success of FWD's several hundred machines already in the sanitary landfill and earth moving business - nothing could compete. This product line alone could thank the efforts of FWD in its continued improvements to make this line a renowned leader, this also applied to the entire product line that FWD manufactured.

Raygo, on the other hand, made an unsuccessful attempt to make a completely different version of the garbage compactor and discontinued the proven FWD models. They also

ABOVE:

> The various ages and sizes of Wagner tractors were suited for all industries from agriculture, forestry and construction. A WA-4 at the front followed by a WA-9 both in early guises of the mid-1950s followed by a Wagner WA-17 fitted with industrial tyres.

attempted to develop a three high container handler, with complete failure.

Shortly after they sold the manufacturing rights on the Wagner Lumber Jack line to Allied Systems; the Chip and Coal Dozer and Piggyback line went to Mr Jack. Raygo kept the compactor line but never produced any of the FWD models. Soon after, the entire manufacturing facility was liquidated and it was only the WA-24 that appeared in the Raygo agricultural line for a very short period of time, in fact only two Raygo Wagner WA-24s are believed to have been produced.

Nearly 60 years on the majority of Wagner tractors are still at work; many have been rebuilt and some re-powered on solid foundations.

Yes, Wagner tractors did have early faults, but the Wagner brothers were honest, continually learning and building on their mistakes, perfecting and expanding their range of tractors, which the rest of the world followed.

The Wagner brothers set the benchmark, they were the pioneers and innovators of the first true 4WD articulated agricultural tractor that we know today.

CLOCKWISE FROM TOP:

> Wagner tractors were used for many industrial purposes from timber through to dirt moving, here a Wagner timber loader, model LJ4-301, the LJ denoting Lumber-Jack, is seen loading an International Harvester semi at White Sulphur Springs, Montana.

> The weight of Wagner tractors was such that it made them ideal leaders in the wheeled sector for earth moving operations. Here a Wagner WS-4 fitted with a FWD Wagner Hydro-Scraper.

> In 1963 this Wagner logger, LG220, was put through its paces before being sent to Africa.

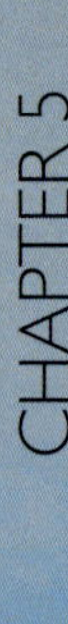

John Deere Wagner WA-14 & WA-17

John Deere described the new line of tractors to join its ranks as;

"They're big – they're powerful – and they're FAST. They're the new John Deere 225hp WA-14 and 280hp WA-17 Four-Wheel-Drive Tractors. Here's a combination of pull power, speed and ease of operation that will delight any large acreage farmer seeking a sure-fire aid to help spread the cost-price squeeze. Either model paves the way for one man to increase the results of his individual efforts."

During the mid 1950s John Deere management and the design team had been looking at ways of increasing tractor horsepower. They had conducted many tests on various tractors including the competition's powerful Case LA powered by a 3-71 General Motors engine and proved that higher power was the only way forward and that there was a growing market for tractors with considerably more power, size and weight.

Wayne H. Worthington, John Deere's director of research was convinced that to increase and utilise more power, the way ahead was with a 4WD tractor and like the Wagner tractor, it should be articulated. The team carried out further tests until they had an experimental 2WD 150hp tractor but it failed in every comparison test against a 4WD articulated Wagner tractor of similar power.

Worthington presented his findings to the various John Deere company directors and invited them to observe several field tests between the two types of tractors. The result was that the directors gave immediate permission and funding for a John Deere built 4WD articulated tractor. By the spring of 1958 the first big John Deere tractor was undergoing field evaluation and was announced to an unsuspecting John Deere two-cylinder following at Marshalltown, Iowa in the fall of 1959.

Production of the 8010 or 'eighty ten' as it was called began in 1960 yet these tractors soon proved too expensive and were described as a tractor too soon. An 8010 cost nearly six times as much as the already successful smaller 4010 tractors. Several problems including transmission troubles occurred with the 8010, every machine was recalled back to Waterloo in 1964. The 84 8010 tractors that were recalled were all updated and returned as the 8020. As many farmers were facing difficulties coping with such a powerful and large machine several models were painted yellow and sold to the construction industry.

John Deere ceased production of big 4WD articulated tractors in 1964; even though they had lost a lot of money on these big machines they had learnt a lot and started designing a new type of 4WD articulated tractor. In the meantime the company could see Steiger and later Versatile beginning to make inroads into the big tractor market and they needed something to fill the gap before their new tractor was ready.

On New Years Eve 1968 John Deere and Wagner signed a deal that entitled John Deere the rights to the WA-14 and WA-17 tractors. Wagner would supply 100 tractors and John Deere would supply the decals, painting the tractors in it's green and yellow colors. The two green John Deere Wagner machines were virtually identical to each other and very similar to the previous yellow tractors, the only visible differences were color, decals, tire size, slight tinwork and chassis modifications. The engines fitted in to these machines were Cummins N855CI engines that were naturally aspired in the WA-14 and turbocharged in the WA-17 model. Several WA-14s were dealer turbocharged at a later date.

Powered by Cummins six-cylinder engines, fitted with Fuller Road Ranger TRO-910 transmissions giving 10 speeds and equipped with air brakes, the John Deere Wagner tractors were ideal for the large wheat farms of the northern states.

The John Deere Wagner agreement only lasted three years with less than the expected 100 tractors being ordered; in all there were 23 JD WA-14s and 28 of the JD WA-17s sold. As part of the marketing agreement between the two companies it was specified that if John Deere stopped buying tractors from Wagner, then Wagner could not produce a competing 4WD articulated tractor for five years. With this exclusive deal between the two companies, basically all yellow Wagner tractor production ended on New Years Eve 1968, total production of Wagner agricultural tractors finished in 1970.

John Deere introduced its new 4WD tractors the 7020 and 7520 in 1971 and 1972 respectively. These new tractors were

ABOVE FROM LEFT:

> Following normal articulation rules the John Deere Wagners had 40 degrees of articulation in each direction and 15 degrees of oscillation from the centre line, the weight distribution was 60% front 40% rear and when in work the weight was spread evenly over each axle.

> Very few John Deere Wagners were fitted without cabs, this particular WA-14 model is undergoing testing by John Deere.

not as heavy as the John Deere Wagners and featured rear linkage and PTO, features never seen before on a 4WD articulated tractor, which appealed to a much wider and larger market of both wheat farmers and corn growers.

When the John Deere Wagner deal was signed in 1968 nearly all the Wagner dealerships closed down. Undeterred, Montana Wagner dealer Willie Hensler along with his workshop foreman Bud Nelson had nothing to sell so went on to develop their own 4WD articulated tractor, the Big Bud. At about the same time brothers Jack and Dave Curtis, who had been heavily involved with Wagner tractor development, went down a similar route developing their own tractors called the Rite, more commonly known as the Rite by Curtis.

These two manufacturers successfully built specialist tractors for the vast rolling prairies of Montana and neighbouring states for several decades to come.

CLOCKWISE FROM TOP:

> A re-decaled John Deere Wagner WA-17 with the FWD decals still working in 2005. The WA-17 was turbocharged rated at 280 engine hp.

> On New Years Eve 1968 John Deere and Wagner signed a deal that entitled John Deere the rights to the WA-14 and WA-17 tractors.

> The John Deere Wagner agreement only lasted three years with less than the expected 100 tractors being ordered.

SPECIFICATIONS

(Specifications and design subject to change without notice)

JOHN DEERE WA-14 and WA-17 FOUR-WHEEL-DRIVE TRACTORS

MODEL . . . . **WA-14** **WA-17**

PERFORMANCE . . . Maximum horsepower at 2,100 engine rpm (factory observed)

	WA-14	WA-17
Flywheel	225	280
Drawbar (4th gear)	178	220

ENGINE . . . . . Cummins 6-cylinder, variable-speed, valve-in-head Diesel

	WA-14	WA-17
Model	N 855-CI	NT 855-CI
Bore and stroke (in.)	5½x6	5½x6
Displacement (cu. in.)	855	855

Governed speed range (loaded) . . . . 2,100 rpm
Operating speed range . . . . 1,500-2,100 rpm
Compression ratio . . . . 14.9 to 1
Lubrication system . . . . Full-flow, full-pressure system, micronic paper element.
Cooling system . . . . 100-gpm pressure system with thermostatic heat control.
Air cleaner . . . . . . Dry type with restriction indicator.

CAPACITIES (U.S. Measures)

Fuel tank . . . . 120 gallons
Cooling system without heater . . . . . . 12½ gallons
Hydraulic system . . . . 52 gallons
Steering reservoir . . . . 40 gallons

TRANSMISSION . . . . . . Sliding-gear type with twin countershafts. 10 speeds forward, 2 reverse, air-assist range selector.

CLUTCH . . . . Foot-operated, spring-loaded; two 14-inch dry disks.

HYDRAULICS . . . . Open-center system. Choice of 15-gpm engine-driven pump and double spool, or 35-gpm power-takeoff-driven pump with either one float and one double-action spool or one float and two double-action spools optional.

STEERING . . . . Full power with two double-acting hydraulic cylinders and individual 30-gpm hydraulic pump.

BRAKES . . . . 4-wheel air brakes and parking brakes.

ELECTRIC SYSTEM . . . 24-volt starting, 12-volt lights and accessories, two 12-volt batteries in series, charged by alternator.

AXLES . . . . 4-wheel planetary reductions in wheel hubs.

TIRES . . . .

WA-14
18.4-34, 8-ply, R-1 duals
18.4-34, 8-ply, R-3 duals
23.1-30, 10-ply, R-1 single
23.1-30, 10-ply, R-3 single
23.1-30, 8-ply, R-1 duals
23.1-30, 8-ply, R-3 duals

WA-17
28.5-26, 10-ply, R-1 single
28.5-26, 10-ply, R-3 single
23.1-30, 8-ply, R-1 duals
23.1-30, 8-ply, R-3 duals

DIMENSIONS	**WA-14**	**WA-17**
Height without cab	8 ft. 3 in.	8 ft. 2 in.
Height with cab	10 ft. 10 in.	10 ft. 9 in.
Width, overall	8 ft. 8 in.	9 ft. 3 in.
Length, overall	21 ft. 5 in.	21 ft. 5 in.
Wheelbase	11 ft. 4 in.	11 ft. 8 in.
Clearance	1 ft. 7 in.	1 ft. 6 in.
Weight without ballast	26,520 lb.	28,280 lb.

GROUND SPEEDS, MPH
(1,500 to 2,100 engine rpm)

Gear	MPH
1st	1.37 - 1.92
2nd	1.75 - 2.45
3rd	2.71 - 3.10
4th	2.80 - 3.92
5th	3.42 - 4.84
6th	4.43 - 6.22
7th	5.61 - 7.88
8th	7.13 - 10.00
9th	8.90 - 12.50
10th	11.00 - 15.50
Rev 1	1.78
Rev 2	5.68

The RITE tractor by Curtis

Following the Wagner John Deere arrangement collapse there was no suitable tractor to fill the gap in the high horsepower 4WD market, two pioneering teams who had been directly involved with Wagner Tractors went in separate directions. Dave Curtis and his brother Jack went one way and developed the Rite tractor while Willie Hensler and Bud Nelson went the other and developed the Big Bud tractor. Both pioneers came from Montana and were manufacturing 4WD tractors only 100 miles apart.

The Rite tractors had the same humble beginnings as Steiger and Versatile and although they never reached high production numbers, tractors were custom built to order for several decades and are still working today.

CLOCKWISE FROM TOP:

> The Rite 750 powered by a Detroit 12V 92T twin turbo diesel producing 750 engine hp could be opened up to nearly 900hp. Weighing in at a little over 67,500lbs (30.131 tons) this tractor was nicknamed 'The Earthquake', its owner said the ground shook like an earthquake when this tractor passed by.

> Taken in 1976, Jack Curtis stands aboard a Rite 404 outside their small factory and workshops.

CHAPTER 6

In 1945 brothers Dave and Jack Curtis started a farm implement business in Dutton, 30 miles north of Great Falls, Montana. Their first franchise was Oliver Equipment and several other short lines, including various brands of cultivators.

Whilst visiting a dealership in Eastern Montana during the spring of 1954, Jack obtained a brochure on the revolutionary Wagner 4WD articulated tractor line, which was manufactured in Portland, Oregon. The Curtis brothers became very interested in the concept of these new production machines and were invited to Portland for a demonstration of the Wagner tractors.

The introduction of the Wagner 4WD tractor range had marked the beginning of the success of the true articulated tractor, and after that Portland visit the Curtis brothers were given one of the first Wagner dealerships in the US and Canada, in their home state of Montana. During that first year they sold 32 of the Wagner agricultural tractors.

Being a little hasty with their sales techniques, the Curtis brothers did not fully appreciate that these first tractors that had been designed for dirt moving in the construction industry were not properly equipped for agriculture. The Wagner tractors were not fitted with planetary axles and were fitted with Waukesha and Buda engines. They were both good engines but not for farm tractor applications. The Curtis brothers, along with Wagner, had a terrific outlay in terms of both labor and capital and something had to be done. Their customers liked the tractors and were very patient, in some cases they helped with the changeover of engines and other major components. However, this experience consumed valuable time that could have been better spent selling tractors. Once the corrections were satisfactorily completed, many more tractors were sold.

The Curtis brothers spent a lot of time and engineering effort helping the Wagner brothers improve their tractors. Dave and Jack became involved with changeovers on the Wagner tractors, installing larger engines and transmissions, correcting driveline misalignment problems caused by the terrible vibrations when turning on the field headlands.

A customer who owned a Wagner tractor suggested that the Curtis brothers' custom build him a tractor of 425hp rather than upgrade a Wagner. They agreed to build their first tractor using many of their own modifications learned from previous problems with the Wagner tractors they had worked on; they used common components of the highest quality throughout in order to keep costs down, with parts and service of these components readily available.

Dave Curtis felt that pioneering the first successful 4WD tractor, the Wagner Brothers in Portland had much to do with the success and dependability of the Rite tractor. The Curtis Brothers had first hand experience of the problems that had occurred in the early Wagner tractors and learned how to solve them.

It was natural to fear the responsibility of a new concept, but after the early experience Curtis felt it was well worth the

CLOWISE FROM TOP:

> The Madson 450 sales leaflet.

> Virtually every Rite tractor can be found still at work today.

effort. Experience is a great teacher; by the time the problems were solved, the solution was deeply imbedded in the mind.

When the brothers built the first Rite tractor, they knew how to match horsepower to components and how to put weight in the tractor to pull a wanted load. It takes weight to pull weight, there is no substitute. Large engines with small axles and transmissions do not work. Curtis felt it was necessary to build on a good base (the axles) and if more power was needed it was comparatively easy to put larger engines and transmissions into the unit. The frame of the tractor and hinge pins were heavy duty on the Rite tractor, the weight ratio between the front and rear units was very important, usually 60% front and 40% rear, the load pulled then equalized the weight over front and rear axles to 50 - 50. This became common in most 4WD tractors that followed.

Not many owners of these big Rite tractors used chloride to ballast the wheels or add other weights on the Rite; the weight was built into the tractor, additional weight was not necessary. Correct size tires were used (pound per square inch) to cut down compaction, with compaction never being a problem with the Rite Tractor.

With experience of alignment problems in other tractors namely the Wagner, the Curtis brothers decided to develop their own oscillation design in the hinge assembly, which would correct any drive line vibration problems. They

ABOVE:

> A Rite 404 covered any tractor in the 400 to 500hp range weighing in excess of 45,000lbs.

designed a drop box with the oscillation around the centre line of the lower output shaft. The end result allowed the tractor to oscillate in any terrain without creating a compound angle in the drive line system.

The Curtis Brothers contracted the Great Falls Iron Works in Montana to build Rite tractors but sadly this agreement quickly ended when the Rite machines were suddenly badged as Madson tractors, with the Great Falls Iron Works calling these tractors their own. It was only after a couple of Madson tractors were built that the correct name Rite came back to the tractor as the Madson tractors were recalled by the Curtis brothers to be repainted and badged correctly.

Three 750hp Rite tractors had been built, one in the Curtis' own frame built by S&S Welding whilst the other two were based on Michigan frames. Ernie Soleberg, a mastermind from Minneapolis, and his company, S&S Welding, had previously manufactured components for the Raygo Corporation. This Rite 750 had a 167 inch wheelbase and weighed approximately 67,000 pounds (29.9 ton) dry.

A tilt cab was designed for easy access to the center components, an Allison six-speed transmission and Cotta two-speed drop box was used.

In 1980 Curtis purchased a Clark 290 M (M for military) tractor unit that was being converted to an agricultural unit. Using just the frame, radiator, Clark four-speed power shift transmission and 70,000-pound axles the brothers mounted a 470hp engine, an Allison five-speed lockup transmission and drive line to the four-speed Clark power shift transmission which gave a 20-speed shuttle forward and reverse.

The dual 30.5 x 32 tires and wheels had already been built and a 60 inch x 60 inch tilt cab was used with all new wiring harness, hydraulic pumps, electrical system, air seat, air conditioning and heater.

The frame on the 290M was one-and-a-quarter inch thick with four-and-a-half inch hinge pins, five inch dual steering rams and a larger radiator made this tractor ideal for large scale farming. The first tractors were doing so well that the Curtis Brothers contacted a company in Augusta, GA who were selling the ex-military units. Jack Curtis flew south to inspect the vehicles, many of which had zero hours on the clock, and ended up purchasing nine units from them.

Curtis used two of the Clark frames to manufacture two 750hp tractors for a large farm north of Fort St. John, B.C. Apart from the axles, the components used were the same as in the previous Rite 750. Dave Curtis, in discussion with a prospective owner, mentioned that the 70,000 pound Clark axles would not stand up to that much horsepower, the buyer said "we won't pull it hard" but the brothers had heard that statement before; the new owner wanted to use the 70,000 pound Clark axles and did not want to spend the money on larger axles. Fitted on 35.5 x 32 tires, during the first season in work when the wheel bearings were stripped, the new owners realized their mistake and asked the Curtis brothers to install Clark 85,000 pound axles the following winter.

Curtis built eight of the Clark framed 470hp tractors with the 20-speed concept. They are all still working today more than 35 years on and are ready for many more years of service. One of the 470 tractors was sold to Saudi Arabia in 1982 and is still running fine in the dry, dusty conditions of the semi arid desert there.

In 1982 the Growth Council in Great Falls wanted to expand manufacturing in the city, they were aware of the Rite tractor

CLOCKWISE FROM TOP:

> Many of these powerful tractors when not out in the fields can be found on huge silage clamps pushing and compacting the harvested crop on large arable livestock farming operations.

> Moving big tractors requires big rigs.

and wanted the Curtis brothers to expand into manufacturing many more units.

At that time they were working double shifts with 20 employees, they did not feel like taking on the responsibility of such a project with management and sales. However the president of a local trucking firm wanted to buy into the Curtis brothers' business, at first he wanted to purchase 49% of the stock but he and the Growth Council Management decided that he should have a 51% share.

After much consideration the brothers felt that it could work, but hindsight told them that it was a mistake, as they later found out when they were not included in any decisions of production, sales or management of their company. It was evident that they had been conned and had sold out for minimum reward; they were tractor builders not businessmen.

Both Dave and Jack soon lost interest in the future of the company that they had built up, the new venture had become a corporation that only lasted two years before the bank foreclosed on them. The Curtis brothers had to buy the assets back from the bank but things became a lot better after that experience says Curtis. The brothers not only learnt how to build a machine, they also quickly learnt even more of how to

CLOCKWISE FROM TOP:

> From the very early days of the 4WD articulating tractor that was designed for agriculture, operators found the weight and traction ideal for dirt moving operations and current manufacturers still build and adapt this type of machine for the construction industry.

> After refurbishment many Rite tractors found themselves being fitted with dirt moving equipment.

make wise decisions about business matters and found there was always something new to discover.

Through the production years the Curtis brothers had several large companies wanting to manufacture the Rite Tractor for them, one being Pac Car of Seattle. The royalty agreement was acceptable but the charge against each tractor for a set of new drawings would consume the profits of many machines. Basically, all the drawings of the frame and mounting brackets etc were in the brothers' possession and the tractors were to be manufactured in the Wagner Mining Loader Plant located in Portland, Oregon; a factory Pac Car had previously purchased from the Eddy Wagner Mining Co.

The second company interested in manufacturing the Rite was an English firm by the name of Dependable Fordath. Its branch was located in a suburb of Portland, Oregon. This company was involved in manufacturing robots at that time but was planning to construct a new plant for building the Rite tractors if the agreement was satisfactory to both parties.

At that time the sales of farm machinery, especially tractors, had deteriorated tremendously and several of the large tractor manufacturers were either downsizing or going out of business, eventually being taken over by larger companies. The tractor market to this day has not returned to the level it was at before the downturn of the mid 80s and Rite tractor development never reached the big time.

Many of the Curtis brothers' new tractor buyers have gone out of the farming business and they have, in many cases, offered to sell their tractors back to Dave Curtis. Of course he has been, and still is, interested in purchasing, refurbishing and reselling these big 4WD articulated tractors.

In 2003 Mervin Laughlin from Youngstown in Alberta traveled south to Montana to visit with Dave Curtis to talk about the future of Rite tractors. Mervin had built his own 425hp tractor loosely based on a Massey Ferguson 4900. Mervin told Dave Curtis that he had used a Rite tractor in Canada for a number of years and it was the Rite tractor that inspired him to build his own tractor.

After several hours of discussion Dave Curtis agreed for Mervin Laughlin to build the 36th Rite tractor. Mervin built the tractor in his farm workshop in Alberta and when all testing was completed the tractor was shipped down to the workshops of Brent Jacobson in Great Falls, Montana where Brent sandblasted the tractor off and sprayed the new Rite 2460 450hp machine in the same yellow as previous Rite tractors.

After the success and interest in the new Rite tractor built and tested during 2004, Lloyd Richie of Boaz, Alabama contacted Curtis with a view to commencing construction of Rite tractors once again. Richie had been rebuilding the Michigan 209M military scraper units into agricultural tractors for several years. Hearing that Curtis had used 290M Michigan main frames and axles on several Rite tractors, he could see a possible niche market for such units.

As of April 2015, 93 year old Dave Curtis has been in the 4WD tractor business for 60 years. The Wagner tractor was the first successful 4WD articulated tractor built. Dave and Jack Curtis helped pioneer the first 4WD agricultural tractors with the Wagner brothers in 1954; they continued to assist them for many years in improving the 4WD agricultural tractor design before becoming tractor pioneers in their own right.

- Rite tractor production started in 1973 with the 404 rated at 425 engine hp from a Caterpillar engine, later 404s were to be powered by Cummins KT 450 engines.
- Engine options were Caterpillar, Cummins or Detroit Diesels with either Allison or Fuller transmissions.
- A total of thirty-five Rite tractors have been built to date with three model 750s built between 1980 and 1982.

Model 303 covering any tractor in the 300 hp + range, av shipping weight 16.96 tons - 38,000 lbs
Model 404 covering any tractor in the 400 hp + range, av shipping weight 20.08 tons - 45,000 lbs
Model 505 covering any tractor in the 500 hp + range, av shipping weight 22.32 tons - 50,000 lbs
Model 606 covering any tractor in the 600 hp + range, av shipping weight 24.55 tons - 55,000 lbs
Model 750 covering any tractor in the 750-800 hp range, av shipping weight 30.13 tons - 67,500 lbs

ABOVE:

> Dave Curtis is still in a position, despite his senior years, to continue building big Rite tractors.This is a 460hp tractor that was refurbished in 2006.

CHAPTER 7

Big Bud history

Big Bud tractors grew from a farm tractor dealership that sold Wagner 4WD articulated tractors throughout Montana. The Wagner Equipment Co Sales and Service was heavily involved with selling tractors in Montana whilst also offering its customers service support.

The company was owned by Willie Hensler whose workshop foreman was Bud Nelson and to his friends he was known as Big Bud.

John Deere purchased the build rights to two Wagner agricultural tractor models, the WA-14 and WA-17, in 1968 leaving the Montana tractor dealership without a source of new tractors to sell. The dealership then concentrated on repairing, modifying and re-building older Wagner units, one of which was re-styled, painted white and called a Big Bud.

The Northern Manufacturing Company was formed from the old dealership in 1969 to build a totally new range of tractor called Big Bud. By 1970 the Northern Manufacturing Company turned to major equipment manufacturers such as Clark for axles, Cummins for engines and Fuller for transmission, as well as other companies for major

CLOCKWISE FROM TOP:

> An original Wagner Equipment Co decal to fit on tractors.

> A modified Wagner tractor, the first Big Bud tractor looked similar and utilised many Wagner components.

components. Bud Nelson designed a frame and assembly of the new 4WD tractor and production began.

Tractors were mostly built to the specifications of the individual farmer who would visit the factory periodically to monitor the progress of his new unit. As horsepower sizes kept increasing to meet the requirements of local wheat farmers, Big Bud tractors soon gained a reputation for size, high work rates and reliability.

The first true Big Bud tractor to be built at the dealership was the HN-250 powered by a Cummins NT855 six-cylinder engine rated at 250 gross engine horsepower at 2100 rpm. This tractor was built using many Wagner parts so looked vaguely similar to a Wagner in overall appearance.

Production Big Bud tractors featured a number of innovations that were incorporated into the Big Bud design which included a patented tilt cab and hood for easy access to the engine, transmission and hydraulics. Big Bud performance was backed by ease of maintenance and ready access for major overhaul points that made a strong impression with many farmers. A simple skid system for speedy power train removal was devised which enabled a quick engine or transmission change.

On 1st January 1975 Ronald M. Harmon purchased the Northern Manufacturing Company. Ron put his knowledge to work, initiating major changes in Big Bud production. His purpose and aim was to build and market larger horsepower

CLOCKWISE FROM TOP:

> As horsepower sizes kept increasing to meet the requirements of local wheat farmers, Big Bud tractors soon gained a reputation for size, high work rates and reliability.

> The new series of tractors introduced in 1975 were called the KT Series which ran alongside the expanding HN Series.

> Big Bud tractors were amongst some of the first machines to pull 60 foot plus field cultivators at 6 -7 mph with ease covering over 50 acres an hour.

tractors than were currently available from any other source.

Within a few months a 450hp tractor was introduced, the new KT Series 1 tractor was built to complement the HN Series 1 machines. In the summer of 1976 Northern Manufacturing expanded into a new 30,000 square foot production facility. The old 18,000 square foot plant was converted for use as a sales and service facility for replacement parts as well as the reconditioning of used equipment.

During 1977 a landmark tractor at 525hp, the KT-525 Series 2 was offered to the market. Far from being prototypes, these high horsepower units were put into full production with sales moving outside Montana to include neighbouring states, Canada, Australia and Iran.

From mid-1977 onwards the HN and KT Series 2 tractors were introduced, the operator cab was widened with several cosmetic feature changes for improved operator visibility. In the year 1977, 36 tractors were produced, while 77 tractors were produced during 1978, both sales and power ratings were increasing at a healthy pace. Marketing efforts included the establishment of a dealer network backed by manufacturing representatives.

In January 1978, Northern Manufacturing built the largest agricultural tractor in the world, the 760hp 16V-747 Big Bud. The unit made considerable headlines throughout the agricultural trade press and was a focal point at the 1978

CLOCKWISE FROM TOP:

> Series I Big Bud tractors were in full production from 1970 through to 1977 with both HN and KT Series tractors.

> In the summer of 1976 Northern Manufacturing expanded into a new 30,000 square foot production facility at Havre, Montana.

> By 1974, Northern Manufacturing was producing 12 to 15 units per year in the 300-350 horsepower range. Virtually all of the tractors were sold in Montana with prices for the units averaging $60,000.

> The clean stylish lines of the new Series 2 or 2nd Generation tractors; here a KT-450 rated at 450hp from a Cummins six-cylinder engine is ready to leave the factory.

California Farm Show. While the 747 was built to order, engineering and design was for a production unit

The 747 had lived up to expectations, being used on a sprawling cotton farm near Bakersfield, California. Following successful field tests, the engine had been opened up to its full potential of over 900bhp.

1978 was a busy year for Big Bug tractor production, the 16V-747 marked the beginning of a new series of tractors, the Series 3 or 3rd Generation Big Buds. A new design, new color scheme and powershift transmissions saw the demise of the HN and KT Series in 1978 with the new series having a successful full production year in 1979.

By late 1978 the name Big Bud had become so well known that Northern Manufacturing Company changed its name to Big Bud Tractors Inc. to reflect that trend.

The Series 3 tractors became increasingly standardized with continued design improvements. In late 1979, a new equalized frame was designed which incorporated an innovative equalization system for improved traction in field turns and over varied terrain.

The new skid chassis could accept engines from Cummins, Detroit Diesel and Caterpillar. Powershift transmissions were now standard on all Big Bud tractors, adding further operating efficiency with shift on the go. Factory installed performance monitoring equipment gave the farmer true control of operating efficiency.

On 5th January 1980 Big Bud Tractors Inc. introduced its highest horsepower full production tractor to date, a 650hp power horse, the 650/50 was designed for both the agricultural and construction industry while the Big Bud 665 was designed solely for the construction industry.

The Big Bud 665 was exactly the same as the Big Bud 650/50 except the Big Bud 665 was fitted with a hard nose front grille which was constructed from ½ inch steel plate that caused the tractor to weigh 1,000 pounds more than the 650/50. This offered improved traction whilst the strengthened front end gave greater protection to the power unit and other components. The 650/50 was fitted on dual wheels all round, 33.5 x 32, which gave a 79.2 inch (6 foot 7.2 inches) diameter wheel of 35 inch wide. Fitted to the 665 were 29.5 x 29 20 ply E3 construction tires and construction series wheels adding additional weight of 9,000 pounds. The total additional weight of a Big Bud 665 versus a Big Bud 650/50 was 10,000 pounds.

These highly sophisticated units epitomised Big Bud design philosophy by incorporating high horsepower in true power to weight ratio for optimum performance. Proven, standardized engine and drive train components with ease of access for servicing resulted in a tractor that could bring great efficiencies to farming, covering more acreage in less time, with less fuel, less maintenance and less farm labor hours.

A figure of 100 pounds weight for every one horsepower was and still is the generally accepted ratio to enable the power to be transmitted to the ground efficiently. In the case of the Big Bud 650/50 this translates into 650 x 100 = 65,000 lbs / 29 ton. The shipping weight of the Big Bud 650/50 was 60,000 lbs with a maximum operating weight of 74,000 lbs giving an average weight of 67,000 lbs / 29.9 tons.

During the summer of 1978, Big Bud began using a Twin Disc electronically shifted transmission as standard equipment for its 525hp tractors. The new electronic transmission exhibited shifting problems attributed to Twin Disc's faulty design. A retrofit program was instituted for units in the field as well as for units in house based upon remedies proposed by Twin Disc. Big Bud participated in the retrofit programme as proposed by Twin Disc. The retrofit programme failed and

ABOVE:

> From mid 1977 onwards the HN and KT Series 2 tractors were in production.

Twin Disc had to fully redesign its electronic transmissions; these redesigned transmissions eventually performed satisfactorily.

Concurrent with the transmission shifting problems, Big Bud began experiencing delivery problems. Parts and components of the Twin Disc transmission consisting of the torque converter, shifter box, cables, filters and the main transmission unit were not delivered on time to Big Bud. Twin Disc's guarantees of prompt delivery of missing parts were given, but not met. Big Bud, relying upon the promises of Twin Disc, continued to manufacture tractors that were complete except for Twin Disc's missing parts and components. Tractors continued to be built but were left uncompleted; these tractors became stockpiled at Havre in large numbers.

The entire situation continued to degenerate as Big Bud, without having complete tractors to sell at wholesale, began to experience severe financial difficulties. In July 1979, R D Tannahill was hired as President of Big Bud Tractors Inc. Mr Tannahill came to Big Bud with 40 years of experience in the farm equipment industry, including 14 years at the J I Case Company during the period that Case rose from near bankruptcy to a leading position in the farm and construction equipment industry. What he encountered was a company with no cash flow, a yard full of tractors missing various transmission parts, vendors clamouring for payments that were averaging nearly four months past due and an inadequate management team.

In the fall of 1979 Mr Tannahill closed the entire operation down and laid everyone off with the exception of the foreman, who was then given the task of completing the units in the yard as and when components were delivered.

Next, he began assembling a new management team; Joseph Atkinson was hired as Controller. Mr Atkinson came from a division of Walter Kidde with nearly 10 years of leadership experience. Hugh Crowell joined as Credit Manager, coming from J I Case Company, where he had 10 years experience in credit and sales. Dan Sinclair was hired as Marketing Manager; Mr Sinclair had over 20 years of marketing experience in the agricultural industry, primarily with J I Case Company.

Secondary levels of management were also added and a field-marketing group was formed. This team quickly set about building a new dealer network of strong, independent agricultural dealers located in the high potential areas for high horsepower tractors. Price Waterhouse Inc. became the company's auditors, replacing a small, local accounting firm.

Meanwhile negotiations for a settlement with Twin Disc continued. By the summer of 1980, it was decided to allow the Courts to settle the matter. Twin Disc filed suit against Big Bud for $1,000,000 representing transmissions delivered to Big Bud during 1979. Big Bud had withheld payment because of Twin Disc's late delivery and also because the transmissions had not met Big Bud's specifications.

Big Bud counter claimed in the Court action because the company believed it incurred damages as a result of Twin

CLOCKWISE FROM TOP:

> In January 1978, Northern Manufacturing built the largest agricultural tractor in the world, the 760hp 16V-747 Big Bud.

> Inventory sat at in the yard at Havre in 1979 awaiting transmissions from Twin Disc.

Disc's failure to deliver acceptable units on a timely basis.

This counter claim in addition requested $1,200,000 as reimbursement for loss of some goodwill, plant shutdown and managerial time attributed to the Twin Disc problem. Big Bud further requested $2,000,000 relating to the improper methods Twin Disc utilised in the attempt to force Big Bud to deal for cash only. Further damages were requested by Big Bud in the sum of $20,500,000 as a contingency claim to cover the possibility of potential damages if Twin Disc's activities were to ever result in forcing Big Bud to terminate business. The company requested a further $27,000,000 in punitive damages against Twin Disc. Big Bud lost the case.

The first quarter of 1981 started with very limited production due to cash flow, vendor and general start up problems. By the fall of that year production had returned to reasonable levels, having risen to one tractor per day.

Surging interest rates and the newness of the distribution system saw very slow retail sales at the beginning of the year, however as the year advanced Big Bud saw a surge of activity as purchases were made to take advantage of tax credits and the lower interest rates available on one of the Company's lease plans.

Financing remained a problem throughout 1981, Big Bud looked for a large capital injection to help pay off the bank debt to allow the company the flexibility it needed to become profitable once again. The construction tractor, the heavy duty Big Bud 665 had successfully completed trials in a Montana coal mine; production of the industrial model was planned after minor design improvements had been made. The construction tractor had a high profit potential, however because of cash flow and capacity problems the introduction of this tractor was very slow.

Throughout 1981 Big Bud experienced numerous disruptions of its manufacturing operations due to inadequate stocking levels of raw materials and components. Because of inadequate cash flow, Big Bud had been paying vendors in the 45-60 day range. The vendors had been reluctant to increase their lines of credit to Big Bud as the Company moved to higher levels of production.

During 1982 financial restraints meant that Big Bud tractor production ceased while the Big Bud Parts and Sales centre remained open to continue supplying and servicing existing clients. Big Bud Tractor Inc. faced Chapter 11 Bankruptcy.

Ron Harmon approached the Meissner brothers in 1982 to

CLOCKWISE FROM TOP:

> In 1980 Big Bud Tractors Inc. introduced the 650hp 650/50 along with the 665 which were virtually identical apart from a heavier front grille on the 665, which was designed for the construction industry.

> To try and encourage farmers to buy Big Bud tractors Harmon and his team flew in farmers from across N America to demonstrate the merits of Big Bud tractors.

help him buy the tractor production business out of Chapter 11 Bankruptcy. In 1985 Joe and Paul Meissner, operating as Meissner Brothers Partnership, purchased the Big Bud Tractor Inc. assets from the bankruptcy court in Great Falls, Montana leaving Harmon to be General Manager of the business.

From 1982 through to 1985 Big Bud tractors continued to be sold. The build up of inventory of nearly 80 units had to be sold off to pay the creditors once transmissions had been fitted. It took just nine months to bring the company to its knees; the Twin Disc and mounting financial problems marked the beginning of the end for Big Bud tractor production.

The heavy duty industrial 525/84 built on a Michigan 290M frame was released in 1984 to try to get production back on line and source new markets.

Big Bud Sales and Service continued to operate looking after existing customers. In 1984 J I Case purchased International Harvester, becoming Case IH and Meissner Tractors Inc. took on the Case IH franchise in Havre during early 1985.

The Meissner Brothers Partnership was subsequently incorporated as Meissner Tractors Inc. In 1986 Meissner Tractors Inc. transferred its manufacturing operations to Big Bud Manufacturing Inc. (later named Big Bud Industries Inc).

Meissner Tractors Inc., through its new company called Big Bud Manufacturing Inc. built two new Big Bud tractors during 1986 to try and resurrect its failing tractor production business; the Series 4 or 4th Generation tractors were built to farmer requests.

During 1986/1987 Meissner Tractor Inc. initiated negotiations with the two Indian tribes, the Assiniboine Gros Ventre Indian tribe of the Fort Belknap Reservation and the Chippewa Cree Indian tribe of the Rocky Boy Reservation, to form a joint partnership under Government funding to train the Indian tribes and develop business skills.

That same year Ron Harmon became General Manager of both Big Bud Manufacturing Inc. and Meissner Tractors Inc., managing the business affairs of the two companies. The Meissner brothers took full control of the Big Bud dealership in 1988.

The company now owned by the Meissner Tractor Inc. and the two Indian Tribes continued to trade as Big Bud Manufacturing Inc. The Indians shadowed the Big Bud team learning both management and technical skills. However this arrangement was short lived; as soon as Government funding ceased, so did the Indian participation.

The Series 4 Big Bud tractors were constructed using many components from the inventory Meissners had purchased from Big Bud Manufacturing Inc. Engines and transmissions were more modern high tech components purchased from some of the world's leading manufacturers.

Production of the Series 4 tractors was very slow. Units were being built to order with only a handful built annually from 1986 through to 1990. Production of the Series 4 Big Bud tractors ceased in late 1990 with only 21 Series 4 tractors being produced. After a 21 year period, 518 Big Bud tractors that fitted a market niche for specialist high horsepower tractors had been built.

Big Bud Series 4 tractor production recommenced at a very difficult time as the agricultural economy was facing a tough time with farmers having little capital to reinvest in big machinery. The introduction of minimum and no-till farming systems meant lower machinery requirements on the farm and

ABOVE:

> Most Big Bud tractors were built and adapted to farmers' individual requirements with several painted in custom colors, such as the Bafus Blue Big Bud.

it appeared farmers had not the need or capital to buy big powerful Big Bud tractors any more.

The globalization of the agricultural industry, in particular tractor manufacturing, meant parts and components could be sourced anywhere in the world. The major companies in the industry were importing and exporting components and finished product at an affordable rate to the farmer. Many tractors were being built with a relatively short life span. The more expensive Big Bud tractor had a virtually limitless life span due to each tractor's initial strong build, which was capable of being upgraded whenever necessary at a lower cost than a new unit.

During a period of low income farmers had to budget accordingly. They needed to be able to see a quick return on their capital machinery outlay; if the cost of a tractor could be recouped over a short period, the books balanced. Big Bud tractors could not compete in this type of market, therefore full Big Bud tractor production ceased in 1990.

Those farmers who have farmed with Big Bud tractors have kept them and still value their correct power to weight ratio today, along with the tremendous work rates that can be achieved. Big Bud tractors will remain masters of the land for many years to come.

Ron Harmon left the management team of Meissner Tractors Inc. in 1993 yet continued to sell for the business for two years before starting his own business, Big Equipment Company LLC. Harmon formed the new company in 1995 to specialise in refurbishing and rebuilding big tractors, of which 80% were Big Bud.

Several of the primary people involved with the design and manufacture of Big Bud tractors are still redesigning and upgrading all models of Big Bud models today. The tractors are stripped down to the frame, rewired, plumbed and upgraded where necessary. The finished product is as near to a new tractor as possible.

Meissners took control of the Case IH dealership under the name of Meissner Tractors Inc. whilst a management team headed by Arnie Lalum continued to develop the business. The dealership became one of the northwest's largest dealers selling combine harvester and 4WD tractors.

CLOCKWISE FROM TOP:

> The Series 4 to some degree incorporated all of the best benefits of the three previous series.

> All Series 4 tractors were built to order, Big Bud tractor production ceased in late 1990 with only 21 Series 4 tractors built.

> Series 4 tractors were introduced in 1986, and work rates could not be beaten.

CHAPTER 8

Big Bud tractor design

Bud Nelson's position within Hensler's Wagner Dealership during the 1960s was that of Service Manager. It was one of Bud's jobs throughout this time to make the Wagner tractor more user friendly for their customers in the area. He literally redesigned several Wagner tractors. He fitted large tires and dual wheels to a series of tractors that were supplied fitted on singles. He also modified various fabrication aspects of the cab and tin work.

The various Wagner cabs were widened and modified to give improved visibility; for improved operator comfort he mounted the cabs on rubber bushes. The hood tinwork again was re-fabricated for improved operator visibility.

When the Wagner John Deere deal was announced in 1968 it took every one including Hensler and Nelson by complete surprise as they suddenly had no new product to sell.

Already redesigning and updating current tractors it was only natural progression to build a new tractor. During 1969 Hensler visited with various farmers and companies to see who was interested in Bud's new design and the prospect of having a new brand of tractor. Throughout 1969 Hensler and Nelson built their first tractor, demonstrating it to local farmers to gain and assess public opinion.

Bud Nelson was a mechanic first and a service manager second, throughout his career he had found various parts on tractors that he had worked on were often difficult to fix, however he knew in his own mind how to remedy those problems.

To remove the transmission on a Wagner tractor for instance, the cab had to be unbolted and totally removed from the tractor. The transmission was then unbolted and lifted out. On other makes of tractor, the tractor had to be split to remove the transmission. This lengthy procedure meant down time was a lengthy problem with repairs out in the field often difficult to undertake.

When Nelson started to build his own tractors from surplus Wagner parts Hensler called them Big Bud, a credit to Bud Nelson who was a big guy, often being nicknamed Big Bud Nelson.

When designing his first tractor Bud decided to do things different. He designed and built a removable power train

ABOVE:

> With the demise of the John Deere Wagner deal, the door was wide open for a new big 4WD tractor manufacturer and that was Big Bud.

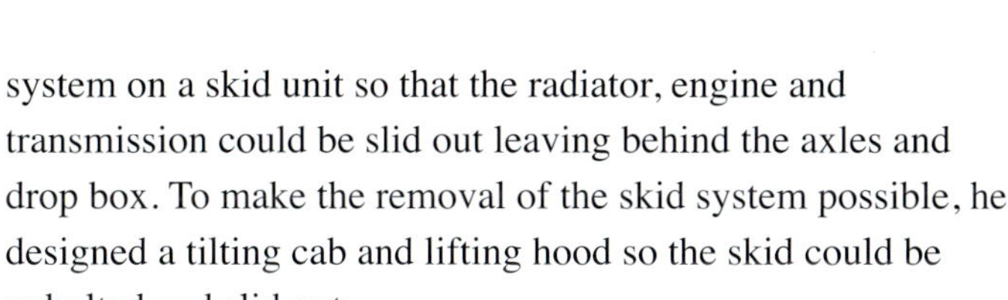

system on a skid unit so that the radiator, engine and transmission could be slid out leaving behind the axles and drop box. To make the removal of the skid system possible, he designed a tilting cab and lifting hood so the skid could be unbolted and slid out.

The removing of the power train skid was possible either in the workshop or out in the field. The skid could then be taken into the workshop where repairs could be carried out and then the whole unit returned to the tractor.

The skid unit was initially designed to carry the Cummins HT 855 series engine and a Fuller transmission. However Nelson incorporated into his unique design, which was an industry first, the ability to replace the older engines with updated models or easily adapt the skid to carry a different make of engine.

Series 1

Hensler and Nelson's first tractor built in 1969 was the HN 250.

The first Big Bud tractor was fitted with a Cummins HN 855 Series engine rated at 250 engine hp. Both Hensler and Nelson used the two letter designation to their advantage, not only did HN denote the use of the dependable Cummins engine but it also denoted the first letters of their names. The number denoted horsepower in the case 250hp.

Powered by a Cummins engine and fitted with a Fuller 12513 transmission, the HN 250 used the same tried and tested components used in the large truck industry. Fitted on Clark 37,500 series axles and 24.5 x 32 tires all round the HN 250 used a Borg Warner one inch by six inch Hyvo Chain drop box system. This chain system in the drop box by changing sprockets allowed for different field speeds to be maintained giving a range of speeds the competition could not match.

The first Big Bud cabs were nothing special apart from the unique hydraulic sideways tilt to enable easy access to the power drive train. A small operator cab 48 inches wide was hardly big enough but it protected the operator from the weather and dusty conditions.

Introduced in 1972 The Series 1 HN 350 had a slightly larger frame with Clark 75,000 series axles and the option of larger 30.5 x 32 tires. The Cummins HN 855 engine was turbocharged and aftercooled to give 350 engine hp with the transmission remaining the same. Physically the two tractors looked identical.

When Ron Harmon purchased the company in 1974 about 20 Big Bud tractors had been built. Harmon's first full production year was 1975 when he took the same tractor concept developed by Hensler and Nelson into which he fitted a Cummins KT 450 engine. Fitted as standard were dual wheels with 30.5 x 32 inside and 24.5 x 32 on the outside with the option to having the larger tires fitted all round. The new series of tractors introduced in 1975 were called the KT Series, which ran alongside the growing HN Series.

By late 1976 two larger tractors were built fitted with Cummins KT 525 engines rated at 525 engine hp. On the larger tractor 30.5 x 32 duals were fitted all round as standard.

1977 was the full production year for the KT Series. The HN 250 was uprated to the 320 engine hp HN 320. By 1977 there were four models in the line up, the HN 320 and 350, the KT 450 and 525.

CLOCKWISE FROM TOP:

> Tractor serial number 6901 was the first Big Bud tractor, the HN250 built in 1969.

> This particular HN 250 serial number 7000 is the first full production Big Bud tractor. The HN250 is powered by a Cummins NT855C six-cylinder diesel engine.

Series 2

The Series 2 Big Buds were introduced in late 1977 with main production commencing in 1978. The Series 2 HN and KT tractors featured a 60 inch wide more spacious operator cab called the Cruiser Cab designed by Keith Richardson. The front of the hood and grille became tapered to allow for increased operator visibility. The hydraulic system was larger and improved to give increased flow rates.

Production moved from the old plant in 1977 to the new larger plant where the newly introduced Series 2 HN and KT tractors were built.

In the fall of 1977 Harmon took an order for a larger tractor of 750 hp plus. Along with his designer and engineer Keith Richardson, they designed and built the giant tractor fitted with a large Twin Disc Powershift transmission and Detroit engine. This tractor was finished and delivered to the Rossi brothers in the spring of 1978, at that time the 16V-747 was the largest agricultural tractor on the market and still holds that title today.

Serial numbers for the early HN and KT tractors could be found on a plate in the cab, with the introduction of the Series 2 tractors the serial number could be found in the cab and also stamped on the upper cross member on the front half of the tractor.

The HN and KT Series tractor chassis were built from half-inch thick steel plate with quarter-inch thick fenders. With the full curved fenders and integrated fuel tanks all the tractors' strength was tied together with the weight spread evenly throughout the unit.

During 1978 Harmon built a prototype tractor for farmer Leo Bitz, who was then the general manager for Big Bud Sales and Service. This tractor was fitted with a Twin Disc transmission; Detroit 8V92 engine rated at 400 engine hp, designated the 400/30. Unlike the HN and KT Series, which still had round fenders, the new tractor had the rounded fenders cut back. This tractor was the first to use the bar type articulation and oscillation system.

CLOCKWISE FROM TOP:

> A 1976 KT450, serial number 7614, rated at 450 engine horsepower from a Cummins 1150KT engine.

> This particular tractor was one of the last of the KT Series built. Serial number 7819 has been uprated to 600hp, hence the KT600 designation, and painted in the new Big Bud livery.

Series 3 or 3rd Generation

The new 400/30 tractor was the prototype tractor for what was to become known as the Series 3 or 3rd Generation Big Bud tractors which featured a different chassis design. This tractor was still constructed from half-inch thick plate, however the reduction of the fender construction slightly altered the chassis strength. From 1979 an all new chassis using one-inch thick steel plate was designed. Leo Bitz's 400/30 fitted half way between a Series 2 and the new Series 3 tractors.

For the new tractor design Big Bud keyed in on the fact that all rear engine mounts on engines built by manufacturers around the world had become standardized. The engine manufacturers, when they designed their engine, did not know what type or make of transmission it would ultimately be coupled to. The mounts on the rear bellhousing were always the same. A useful point when deciding which engines and transmissions would be used in a new design of tractor.

Big Bud bolted its engine mounts to the side rails rather than welding, as the opposition had done. This way it left the option that an updated engine could easily be fitted to the tractor at a later date. The width between the mainframe rails was over 48 inch giving ample room for any current or future engine between 400 and 800hp from any manufacturer to be able to easily fit. The front engine mounts became the mounts for the radiator, so if a replacement engine of a different length was fitted, the radiator was just moved back or forwards depending on the overall engine length.

After the Twin Disc troubles where Big Bud had over 70 new tractors stood waiting for transmissions to be fitted, a universal mounting system allowed for the fitting of virtually any type of transmission. The Series 3 Big Buds were designed around a system that still allowed for instant access and was a step forward in addition to the removable skid system, the cab and hood were still hydraulically operated to tilt and lift. The system never used the overall strength of the components coupled together. The frame was fully self-supporting.

Many competitors at that time, and still today, design and construct a tractor around the engine and transmission, which becomes part of the tractor's overall strength. When a component has to be removed for whatever reason, the tractor has to be split involving considerable down time.

The one-inch thick frame rails of the Big Bud tractors were pre-drilled at the time of manufacture to accommodate any make of engine or transmission that could quickly and easily be bolted into place at a later date. The basic idea of this type of system was ease of maintenance and longevity of product. The main steel work was strong and reliable, so instead of scrapping a tractor when the engine and or transmission were finished, why not totally refurbish it fitting a new modern engine and transmission. These changes were introduced from the introduction of the 360/30 in 1979. The removable skid power train system of the earlier models was no longer required.

The HN and KT Series Big Bud tractors were take-offs of Bud Nelson's original design, from 1979 Harmon and his team completely redesigned the way the tractors were built.

It was now possible to have a choice of Cummins, Caterpillar or Detroit engine fitted to the tractor. The range of transmissions tried were Twin Disc, Allison and Clark.

For the new Series 3 tractors the Twin Disc full Powershift transmission with torque converter proved the best for the type of tractor and the variation of work carried out. The Twin Disc transmission had quite a unique feature, not pre-set by the manufacturer it was possible to adjust the point when it came in and out of lock-up in all speeds. The advantage being a

ABOVE:

> Serial number 7919, the 1979 450/20 is hooked up to a Friggstad 60 foot chisel plow with 60 foot John Deere grain drill capable of working over 40 acres an hour.

nearly infinite variable speed transmission, gaining maximum use out of all speeds from 0 to max speed in each gear. A tractor such as the 750hp 747 could work at one mile per hour all day without transmission overheating, giving full utilisation of horsepower. Other transmissions which were standard within the industry; the point of lockout was pre-set. A Big Bud could achieve maximum torque maintaining speed in the most favourable gear whether going up hill or through tough spots.

Series 3 tractors continued to use Cruiser cabs and Clark 7000 series axles. Cummins KT 1150 engines were used for the 525 and 600hp tractors. Smaller tractors such as the 320/10 used the Cummins 855 engine. Caterpillar 528 or Clark 37,500 pound axles were offered as an option.

Designing of the Series 3 tractors was very much a team effort. Farmers and tractor drivers were consulted every step of the way because they knew what they wanted and what they required from their tractors in the future.

College trained engineers were ideally suited for many design features, but lacked the practical experience of the operator out in the field. Ron Harmon's close friend Keith Richardson, a man who had worked many hours on a tractor and who was a self taught engineer, was the main design engineer. Every prototype and production tractor Keith built is still running today proving the philosophy that tractor design should be a team effort between operator and engineer with Richardson fulfilling both criteria.

The tractor color of the Series 3 Big Bud remained Imron White, a color farmers across America had come to recognise. The red decals highlighted in black were the standard, but various custom built tractors appeared in various other decal colors.

Series 4 or 4th Generation

When production of the Big Bud tractors recommenced in 1986 a totally new tractor was designed and marketed. The look of the tractor and the use of electronic transmissions and high tech engines bought Big Bud tractors up to date with the majors who were beginning to make greater use of electronic computer controlled engines.

The ROPS cab was retained from the earlier models, as were the Clark 70,000 pound series axles. A redesigned main-frame and articulation system improved traction and overall operation.

One of the main modifications the management team wanted to accomplish was to reduce wear and potential failure around the drive-line. When the tractor articulated and oscillated at the same time, usually on headland turns, drive-line vibration problems occurred causing unwanted wear.

The tinwork was cosmetically improved and tapered; this was done to allow the operator maximum visibility and, along

CLOCKWISE FROM TOP:

> This 1979 525/50, serial number 7916, pulling a 59 foot air seeder is heading home after autumn seeding. The Big Bud 525/50 tractor models quickly gained the name of a Big Bud Five and a Quarter.

> There were only two Big Bud Series 4 tractors ever to be factory-built on triple wheels. This particular 440 was built to order in 1987.

> All Series 4 tractors were built to order, as was serial number 90507. This particular Big Bud 450 went on to a 3,900 acre farm, it was built in the fall of 1990 and first used in the spring of 1991.

with the new paintwork and decals, the whole tractor looked more modern and aesthetically pleasing.

The tractor hood now raised electronically, when it reached its maximum lift it was then automatically possible to electronically tilt the operator cabin back, power train components were now far easier to reach.

The design team wanted to make the whole new system stronger and easier to upgrade with improved accessibility to all the major components, they went a step further to make routine maintenance and servicing easier for both the operator and mechanic.

The Series 4 to some degree incorporated all of the best benefits of the three previous series. The whole unit was once again even simpler to upgrade, designed to take any make of engine and transmission.

Batteries were fitted in the front of the frame, the radiator and grille were mounted on the front of the engine, the air intake system now became part of the power train. A mere six bolts held the engine and transmission in place, when the bolts were removed the whole power train could be lifted out and taken into the workshop. While the whole unit was on the workshop floor changes and adaptations could be carried out and tested without having to fit the power train back into the tractor.

The Series 4 370 through to the 500 Big Bud models were constructed from a one inch frame and fitted with Fujitech electronic transmissions as used in the heavy duty version of the Case IH STX tractors. This type of transmission had in the region of a 500hp limitation.

The two larger 740 tractors used a one-and-a-half inch thick steel chassis. The transmission was a 2610 Twin Disc power transmission with torque converter. The Komatsu 170 series engines produced 740 engine hp. Komatsu axles were used and the whole tractor was mounted on 35.5 x 32 duals all round.

All Series 4 tractors were built to order, in total there were only 21 units built of this last Series with Big Bud tractor production ceasing in late 1990.

ABOVE:

> Only two of the Big Bud 740s were ever built. The 1987 model 740 went to Willow Brook farms in Florida, the previous owner of the Big Bud 747, whilst the 1989 model 740 went to the Warden Hutterian Brethren of Warden, Washington and was sold in 2003 to Progress Farms Bakersfield, California.

CHAPTER 9

Big Bud manufacturing policy

The manufacturing policy of the company was to purchase major components such as engines, transmissions, axles and hydraulics and to fabricate the tractor around these components. Thus the actual manufacturing operations of the company would include cutting, welding and shaping the steel and the actual assembly operations. There then was no need for heavy manufacturing.

The first advantage of this type of manufacturing strategy was that it allowed Big Bud to achieve extremely high standards of quality. The company was able to search through brand name manufacturers such as Cummins, Detroit Diesel, Fuller and Clark to select standardized components that met Big Bud's design requirements for performance and reliability.

The major competitors in the industry might have had a reasonable reputation for overall quality, but none of them had the expertise needed to match the quality levels achieved by the speciality manufacturers across the whole spectrum of components.

The second advantage of this philosophy was that it allowed the company to carry very low fixed costs due to the low levels of capital goods required to support this type of manufacturing policy. In the event of an economic downturn, lower volumes did not translate into high unabsorbed burden costs. Furthermore, in periods of economic expansion, the capacity was normally available at these vendors, or a secondary vendor could have been approached to meet the volumes required. The end result was a very short lead time.

The third advantage was that it allowed the company to

ABOVE:

> From the early days of Big Bud tractor production it was the company's philosophy to over engineer, build strong and build in reliable components to satisfy the farmers every need for a big tractor.

spend far less on research and development. Big Bud's research and development concentrated largely on chassis and modular design as well as improved product features.

Money was not spent on research and development in order to achieve a fuel efficient engine, a Powershift transmission or any other features that were standard on Big Bud tractors. The dollars of research and development needed for these types of activities was spread across high volumes of production by the speciality vendors such as Cummins and Fuller.

The fourth advantage of this type of philosophy was that the company could compete in very low volume markets. The major competitors in the industry were not interested in entering the markets that Big Bud operated in because the low volumes could never justify the capital expenditure that would be required to produce the products serving these markets.

Major manufacturers wanted to produce as much as possible in house for a variety of reasons. These reasons were typically the disadvantages Big Bud faced by following their philosophy. The end result was, however, that Big Bud was able to avoid direct competition with the majors by seeking niches in the market that, because of lower volumes, the majors were unable to serve.

The fifth advantage was serviceability. The end user did not have to rely solely on Big Bud for servicing his product. Because of the universality of the components used, there was generally service available somewhere in the end users' geographic area.

An example of this would be a Cummins engine that was used not only in Big Bud products but was also used on numerous other pieces of construction, agricultural or trucking equipment. Clark axles were also used on a variety of applications. Thus, any of those companies' distributors would be able to service those particular components on the products that Big Bud manufactured.

CLOCKWISE FROM TOP:

> Easy access to all components was a key factor to marketing Big Bud tractors.

> The production facility at Havre was a relatively small operation where steel cutting and fabrication was the core business.

> Steel fabrication equipment and cutting equipment at the small factory where assembly lines did not exist as would be the case with a large tractor manufacturer.

> A skid unit being slid into place into a 1990 Big Bud 440.

This manufacturing philosophy did have its disadvantages. The main disadvantage was the ability to control one's costs since the manufacturer utilizing this philosophy was reliant upon its suppliers. Obviously, in the company's history pointing to Twin Disc and the problems Big Bud had encountered with them pinpoint the disadvantages of this philosophy. Furthermore, cost increases could be passed along and the tractor manufacturer had little option but to accept them.

Big Bud had been very concerned with these problems and as a solution developed at least three vendors for every major purchased component.

Additionally as the company grew it exerted more and more leverage over its suppliers. Big Bud was already at the point where for each of the three or four major suppliers it constituted four to five million dollars worth of business annually. Also cost disadvantages on the major components that the company purchased were generally offset by higher margins on the products sold because of the lack of direct competition from the major competitors in the industry.

Big Bud had another advantage over the large competitors in the industry in that the company was non-union. This translated into increased flexibility in terms of working schedules, work assignments and numerous other factors. Further more, the wage rates in north central Montana were considerably less than manufacturing averages throughout the United States, especially when compared to the average U.A.W. wage rate that most major manufacturers in the industry were faced with.

An inherent part of Big Bud's manufacturing philosophy was product quality. Big Bud owners placed a very heavy degree of reliance on their tractor since, in many cases, they had traded their smaller tractors when buying the Big Bud and

CLOCKWISE FROM TOP:

> Due to the sheer size and weight of the tractors and dual wheels, tractors were often loaded in the workshop where overhead cranes could be used.

> An impressive sight as a Big Bud with its dual wheels loaded on the trailer heads out to another customer.

> Many Big Bud tractors traded in are totally refurbished and leave the facility in as-new condition.

> Special permits were required in various States to allow such big loads to travel along the highway.

therefore did not have the backup they formerly had available. The company's concern for quality was evident on closer examination of the tractors. Big Bud tractors were over-engineered with every component designed to take far more than could reasonably be expected to occur in normal activity. Even so, there was always the possibility of a failure despite all the precautions one may take. Big Bud had sought to counteract this possibility in the design of the product.

Ease of serviceability was evident throughout the tractor. Every component was easy to reach and remove. For instance, the entire hood tilted completely forward, the cab tilted back in the opposite direction and any of the major drive train components could be easily repaired, removed or replaced in the field.

Full time factory service representatives were always available to assist when needed, but they were rarely needed because of the high quality of the distribution system that the company had. Each Big Bud dealer was large, well capitalised and respected in its area for the quality of its service work. When Big Bud selected its dealer network these were the prime criteria considered.

Fortunately this issue of quality had been of paramount importance since the very early days of the company. As a result, Big Bud already had many satisfied customers that were willing to testify to the tractors' reputation for quality. Once a company lost this reputation for quality, it was rarely, if ever achieved again. Big Bud did not intend to lose its quality reputation because only by competing with quality could any company achieve the margins that allowed for a profitable performance.

CLOCKWISE FROM TOP:

> Skilled engineers are employed to work on and refurbish big tractors leaving the Big Equipment Company.

> For easy access the cab and the hood both tilt in opposite directions. Series I KT-450 clearly showing the turbocharged Cummins big cam engine.

> Access to all main components, whether engine, transmission or electrical, is an easy operation either in the workshop or out in the field. The hood and the cab tilt with ease through an electro-hydraulic system.

BIG BUD UPDATE

Big Bud is again successful. On October 16, 1981, the company successfully met its payment schedule to the Seattle First National Bank and First Security Bank of Havre, representing a total pay down and take-out of Six Million Dollars since July 17, 1981. Immediate plans call for continuing manufacture of tractors.

As part of the pay-out and take-out of the Banks, the Banks have agreed to become part of new funding and the Banks will fund operations during Big Bud's remaining restructuring, including the sales of Big Bud stock. Big Bud is offering stock to Montana residents upon conditions set by the State of Montana Securities Commissioner. Response to the outlook for continued sales of stock along with Big Bud's increased marketing in the United States and World-Wide is excellent. In the ninety days, during which Six Million Dollars of the Bank debt was repaid, Big Bud had its highest sales in the company's history for the months of July and August.

The management team comprised of Ron Harmon, James Massick, and Mort Goldstein attests its succcess to the outstanding reputation of Big Bud's tractors and the continued increase in demand for the Big Bud product in the agriculture, mining, and construction industries. Production of the large horsepower models is continuing with design plans for construction of additional smaller 4-wheel drive tractors for International markets. The Big Bud-Friggstad scraper combination is now in operation at the mining area at Colstrip, Montana.

Big Bud Tractors, Inc.

P.O. BOX 1111 • HAVRE, MONTANA 59501
PHONE 406-265-5457
TLX 31-9558

October 21, 1981

TO: All Dealers and Employees

RE: Big Bud Tractors, Inc.

It is with great pleasure that we announce the completion of the bank agreement that matured on October 15, 1981. This completes another step in the re-organization of our company. This event takes out the major time emergency faced by the company and allows us to proceed in an orderly fashion to complete the raising of equity capital for the future growth and development of the high horsepower, high quality tractor business.

Your assistance, understanding, and patience during the past several months has been a major factor in this accomplishment. We are now in a position to ship, floor plan, and actively promote the sales of the world's best tractor to cost and quality conscious tractor customers. Prospects, dealers and employees can face the future with confidence.

Immediate plans call for the continuing manufacture of tractors. The first seven tractors scheduled are five (5) 525/50's and two (2) 650/50 tractors. We will gradually increase production over the next several months and should be back to full production by February.

Based on your retail forecasts we are looking forward to excellent retail activity during the balance of 1981.

LET'S GO TO WORK!!

Very truly yours,

Daniel G Sinclair

Daniel G. Sinclair
V.P. Marketing

CHAPTER 10

Exporting Big Bud Tractors

Once the Big Bud tractor had become established on the vast wheat growing farms of North America an export plan was put into action. Potential markets overseas were studied with a plan implemented to export tractors to Australia, Iran and the Middle East, the Philippines and the larger European farms.

From the selected potential overseas markets a small handful of tractors were exported to Iran and the Philippines with the largest sales achieved in Australia.

The Australian agricultural landscape was like nowhere else on earth, farms were far larger than those found in North America with farmers working in virtual isolation for long hours at a time.

The component built tractors that were easily serviced, maintained and repaired were ideal for this market.

Australia

During the fall of 1980 the Australian press were having a field day, the American Big Bud tractors had arrived. The importers, Horwood Bagshaw, believed these big tractors would bring big savings in time, labor and fuel consumption; this had been demonstrated conclusively in Australia.

A Queensland farmer had found that his 525hp model would rip the same amount of land as two D7 Caterpillar tractors using less fuel. Working side by side the Big Bud just left the D7s standing.

This was not the first attempt by Ron Harmon to export Big Bud tractors to Australia, in fact he had been fairly successful on several occasions but his biggest let down was not the farmers themselves as they needed big powerful tractors. The problem was the dealer network; Ron Harmon was finding it difficult for someone with a big enough network to handle his tractors.

Several Big Buds including Series 1 and Series 2 HN, KT and Series 3 models were sold to work in Western Australia around Perth, in Queensland and New South Wales competing with Australian built tractors like the Acremaster, Baldwin and Waltanna.

The cost of shipping big tractors around the world was an expensive exercise for Big Bud Inc, the giants, John Deere, Steiger and Versatile were in a much better position to supply big 4WD tractors to the lucrative market in Australia.

Over a ten-year period Big Bud tractors were exported to Canada, Iran, Kenya, Australia and the Philippines. Had Big Bud continued in business no doubt the name and reputation would have travelled around the world and many more Big Bud tractors would have been exported, especially to Australia.

CLOCKWISE FROM TOP:

> During the fall of 1980 the Australian press were having a field day, the American Big Bud tractors had arrived. Farmers flocked from all over.

> Ron Harmon aboard a Big Bud KT-450 in front of a group of enthused Australian farmers.

Press Release 18 November 1980
Horwood Bagshaw Ltd Edwardstown
Southern Australia.

The biggest of the big tractors

'The biggest agricultural tractor so far available in Australia was launched officially in Brisbane today.

With the unveiling of a U.S. made 650hp, 33 tonne machine, Horwood Bagshaw Ltd, launched the Big Bud Model 650/50 on the Australian market. But the 650 is not the biggest tractor made by Big Bud Tractors Inc., of Havre, Montana. A 760hp, 59 tonne giant – the Model 747 - will soon be available through Australian distributors, Horwood Bagshaw Ltd., and development of a 1000hp machine is at an advanced stage.

With a range of 400, 525, 650 and 760hp tractors, which sell at between $142,00 and $446,500 (Australian dollars), Big Bud tractors fit into the top sector of the rapidly growing four-wheel drive tractor market. The giant tractors allow farmers and miners to pull bigger implements using less fuel and labour, and in a shorter time – resulting in significantly lower production costs. Big four-wheel drive tractors are no longer a novelty in Australia. Big Bud has established itself world-wide as a leader in big tractor technology, with careful design and development resulting in higher field efficiency than other four-wheel drives.

Features and options for the tractors include such things as fully air conditioned cabs with full instrumentation, mini computer monitoring for efficiency and early warning of potential problems and closed circuit television monitoring of implements being pulled.

Since taking over the Big Bud Franchise in June this year, Horwood Bagshaw has found an encouraging enquiry for the tractors, particularly in the top section of the range.

ABOVE:

> Australian agricultural landscape was like nowhere else on earth, farms were far larger than those found in North America.

Big savings in time labor and fuel

Big Bud believes that its tractors can bring big savings in time, labor and fuel costs. This has already been demonstrated quite conclusively in Australia.

One farmer in Queensland has found that his Big Bud 525 will rip the same amount of country as two D7 Caterpillar tractors. Working side by side the 525 used 1.4 gallons of fuel an acre compared with 2 gallons an acre for the D7. Applied over an area of 2,000 acres this represents a fuel saving of 1200 gallons of fuel or about $1,560 (Australian dollars) a working costed at $1.30 a gallon.

Another farmer who was previously able to blade plow three acres an hour using his D8 found he could cover seven acres an hour with his new 525 Big Bud. Fuel consumption for the D8 was 3.66 gallons an acre compared with three gallons an acre for the 525. Calculated over an area of 3,000 acres this represents a fuel saving of 1,980 gallons or $2,574 in cost.

In addition he saved 572 hours working time with the Big Bud. Costing labor at $6 an hour this represents an additional saving of $3,432.

Big Bud believes that big savings can also be achieved by replacing multiple small tractor operations with large single units.

The company quotes the case of a farmer in the U.S. who replaced two 300hp tractors with one 525hp Big Bud.

Pulling an 18 furrow plow under average conditions, the 525 covered 15.1 acres an hour compared with a combined performance of 12.54 acres an hour for the two 300 hp units, each coupled to a nine furrow plow. The 525 took 199 hours to work 3,000 acres, compared with 478 hours for the two smaller units.

Using 16.7 gallons of fuel an hour, the 525 needed 3,323 gallons to work the 3,000 acres, compared with 5,593 gallons of fuel for the two smaller units using 11.7 gallons each an hour over the same area. Using an average fuel cost of $1.50 per gallon (projected over seven years), the 525 saved the farmer $3,405 in fuel each time the land was worked. Assuming the land was only worked twice each year the saving in fuel costs at the end of the seven years would be $47,677 according to Big Bud.

ABOVE:

> Field tests of Big Bud tractors in Australia proved savings could be made in all areas from work rates, man hours and fuel costs.

In addition the farmer also saved 239 hours of labor costs with each working by eliminating one tractor. Computed over seven years with two workings a year and an average labor cost of $6 an hour, this represents a saving of $23,422.

Ease of maintenance is another big selling point for Big Bud. Features include self cleaning front grille to avoid chaff build up, easy access to air conditioner and battery, tilting bonnet and cab for easy access to engine, transmission and hydraulics and preventative maintenance features such as extensive warning gauges and optional Sentinel shut-down system and computer monitoring for tractor functions.

Other factors also need to be considered, such as long life engines (e.g. 10 – 12,000 hour life expectancy for the Cummins KTA 1150C in the 525/50), 1 inch (25mm) and 1½ inch (37mm) thick steel plate frames for rigidity and solid attachment points, over specified transmissions and drivelines and the low slippage factors which result in greatly decreased tire wear.

Ease of servicing a major feature

The design of Big Bud tractors leans heavily towards ease of service and maintenance. For example the extremely heavy (up to 1½ inch 37mm section) chassis and frame is totally self supporting and does not rely on the engine and transmission housings for strength.

Components like the cab, grille and radiator, are hinged so that they can be quickly swung out of the way to give free access. This means that any of the mechanical components can be lifted from the tractor without having to split the chassis.

The engine mounting system allows the power plant to be slid out from the front of the tractor with only minimum uncoupling of components. The cab is equipped with its own hand operated hydraulic jacking system for raising and lowering it in the field.

A major feature of the self-supporting frame design is that it allows for component interchange and upgrading if required at a later date. Provision is also made on the engine and transmission mounts for different makes or models of power plants and gearboxes to be fitted. The same provision is also made for axle replacement or updating.

By adopting this design philosophy, Big Bud has been able to minimise down time caused by servicing and repairs. It also means that in many instances repairs can be done in the field without sophisticated equipment.

CLOCKWISE FROM TOP:

> The component built tractors that were easily serviced, maintained and repaired were ideal for the Australian market.

> The Australian Press declared on 18th November 1980 that the world's biggest production line farm tractor, the Big Bud 650/50, made its Australian debut in Brisbane. The 650hp tractor would sell for $250,000 Australian Dollars with Horwood Bagshaw, the South Australian company, handling sales of the Big Bud.

> The export market for Big Bud tractors in Australia saw several model series introduced and sadly even though they proved their worth they could not compete with major tractor manufacturers from North America.

Keller's Supply & Service
9630 Packard Rd.
Morenci, MI 49256
Ph. (517) 458-6312

FOUR WHEEL DRIVE CUSTOMERS:

ATTENTION:

Keller's Supply Inc. are pleased to announce that they have added to their line BIG BUD Tractors to fulfill your 4-wheel drive-high horsepower needs. With more acres and less time to cover them, BIG BUD'S 400 plus horsepower can get the job done. BUG BUD will be arriving at Keller's Supply October 14, 1980. Keller's invites you to a field demonstration at their Dealership on October 14th. You will see and drive BIG BUD to prove that BIG BUD has No Excuses.

You will also see demonstrated a 19 tooth 24' Glencoe Soil Saver and a 12 bottom Will Rich Plow. Come see NO EXCUSES BIG BUD all day October 14th at Keller's Supply. Rain date October 16th.
BIG BUD is built for reliability!

NO Excuse
BIG BUD
Keller's Supply Inc.
Daniel R. Keller

Big Bud Tractors, Inc.

P.O. BOX 1111 • HAVRE, MONTANA 59501
PHONE 406-265-5457
TLX 31-9558

November 6, 1981

Kellers Supply & Service
9630 Packard Rd.
Morenci, MI 49256

Dear Dan:

RE: Sales

On 8/18/81 you submitted the attached forecast of tractor sales for the balance of 1981.

Historically 40% of the annual four wheel drive sales in the U.S. are closed in the last quarter of any given year. Based on the (FIEI) Farm and Industrial Equipment Institute numbers 1981 is no different than any other year. As of the current report four wheel drive sales are up 2.8% year to date over last year with approximately 60% of the 1980 sales completed so far this year. This means there will be between 4300 and 4500 four wheelers sold and settled during the last quarter of 1981.

Retail sales are the life blood of any company. You can help your dealership's present and future financial outlook by completing your forecast at an early date. Let's get our share of this market while people are in the buying mood.

Will you help us get in on the action and meet your forecast?

SELL BIG BUDS!!

Yours truly,

H.E. Crowell

H.E. Crowell
General Sales Manager

CHAPTER 11

The Big Bud 16V 747

Big Bud 747

It is in north west Montana where you will find the largest agricultural tractor in the world hard at work. Here there is an area of prairie countryside known as the Golden Triangle. Starting at Great Falls head north west to Shelby then east to Havre and back south to Great Falls and this is Big Bud country where rainfall is usually 11 inches per annum and near dust bowl conditions often prevail.

About half way back down to Great Falls on Highway 87 you come across a small town called Big Sandy, head east and this is home to the Big Bud 16V 747. After Big Sandy good luck because the dirt roads just go on for mile after mile with nothing to see but beautiful scenery, dust, wheat and if you are lucky enough, monster tractors.

Big Sandy lies in the shadow of the Bear Paw Mountains, where you will find the Williams Brothers, Robert and Randy farming in the region of 8,000 acres of wheat and garbanzo beans. The original Williams family first moved to this isolated but majestically scenic area, home to the Bear Paw Indians, back in 1913. The grandparents arrived as settlers in the area with nothing of their own, they were given a piece of land neighbouring their present spread where they set up ranching.

By 1944 Robert and Randy's parents had acquired the present homestead where they farmed in the region of 700 acres of cereals and livestock.

As the farm grew in size and the brothers became old enough to work and manage the land themselves, the machinery on the farm started to become more modern and more powerful. These two young men were power, tractor, car and motorbike mad, all the time wanting bigger and more powerful machines.

By the late 1970s the biggest tractor they had on their farm was a Case 1200, the first tractor built by J I Case to incorporate four-wheel steer, 4WD and with its six-cylinder turbocharged diesel engine rated at 120hp work rates increased at such an alarming rate that the brothers were in a position to acquire and farm more land.

It was in 1977 whilst out motorbike riding with their good friend Ron Harmon, owner of the Northern Manufacturing Company which was building Big Bud tractors at that time, when they heard of the plans to design a big tractor, probably the biggest agricultural tractor ever to be built.

The brothers would spend hours at the Big Bud manufacturing facility in Havre watching a pile of metal and the big Detroit engine evolve to become a monster power horse weighing over 52 tons. At that time neither Robert nor Randy, who were in their early 20s, ever imagined this tractor would one day become theirs and return to work less than 50 miles from where it was built.

Building of the Big Bud 16V 747 was started in 1977 and completed by January 1978 ready for shipping south to its new owners the Rossi brothers. It was the world's largest and most powerful tractor.

Big Bud tractors, during the mid 1970s, were increasing in

ABOVE:

> Looking at tractor owner operator Robert Williams in the cab of the 747 gives some idea of the scale of the world's largest agricultural tractor. An impressive sight at 20 foot wide and 13 foot 6 inch to the top of the cab.

horsepower by as much as 100hp at a time to meet the demand for increased work rates, a decrease in labor and the use of larger implements on the land. The KT Series reached 525-engine hp on tractors commonly called the Big Bud five and a quarter.

Ron Harmon felt that a 525hp tractor was too powerful for a mechanical shift transmission; clutches were not strong enough and many clutching problems started to emerge as power increased.

"I wanted to move away from a standard shift transmission to a powershift," stated Ron. "I felt 450hp was the maximum we could reach with a standard shift transmission with no clutch problems, we had plans to build tractors much more powerful with an eventual aim of reaching 1000 engine hp so a powershift transmission was essential."

"At that time the Rossi Brothers from Bakersfield California had been using a Big Bud 525 and liked it tremendously. They had become good friends of mine over the years and when I mentioned I was planning a new series of Buds called the Series 3 and the power would be increased, they immediately wanted one of these tractors with a power band of between 750 and 1000 engine hp" added Harmon. "We worked closely together planning and designing this new tractor and it was agreed that the Rossi Brothers would have this first tractor as a prototype and I would be able to monitor and assess its performance out in the field. This way they got a big tractor for their farm for approximately $300,000 and I would be able to build and test the largest agricultural tractor currently available."

The Big Bud 16V 747 was designed as a production tractor, but only one was ever built. It was mid-1976 when the idea was put into motion for this tractor, the Rossi Brothers spent a lot of time evaluating their needs and requirements, building of the tractor started in mid 1977.

Ron met with a leading manufacturer of transmissions, Twin Disc, which was building systems for some of the largest construction machines mainly in the coal industry and between them they came up with a transmission to suit. A full powershift Twin Disc TD-61-2609 with six forward and one reverse speed was chosen to move this monster at a top speed of 20 mph. The powershift transmission allowed for the smoothest possible delivery of power to the axles while giving the operator the ability to make use of the best engine torque range over a wide variety of field speeds.

Next came the engine, it was decided to use a Detroit Diesel 16 cylinder engine in V configuration as this had a minimum rating of approximately 760hp and could be opened up to over 1000hp if ever the need arose.

No attention to detail had been spared in the design of the machine or cab. Many features were incorporated into the cab design to make the long days at the wheel more pleasurable. Full air conditioning and heater allowed the operator to work around the clock if need be. A surround sound stereo system and 40 band CB radio along with a closed circuit TV to monitor implement performance helped stop the driver from falling asleep.

Power steering was essential, a decelerator pedal for end of field turns made life easier and full instrumentation to warn the operator of any possible fault put this tractor way ahead of the competition at that time.

CLOCKWISE FROM TOP:

> There was no production line when building the 747, the shell sat on the floor while everyone worked around it.

> The 747 was designed as a production model but only one was ever built, build started in 1977 and was finished in January 1978, a mammoth task in its own right.

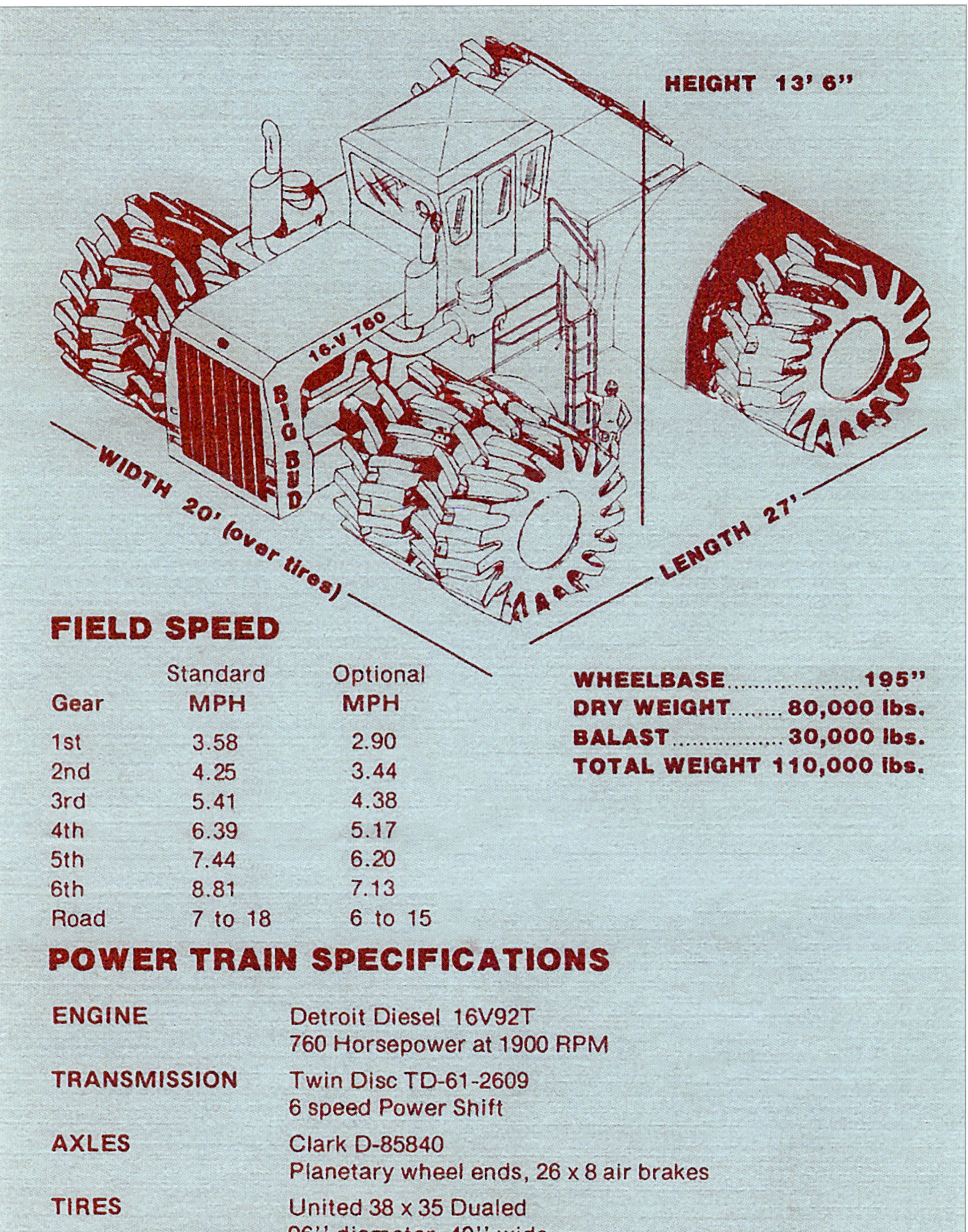

WHEELBASE .... **195"**
DRY WEIGHT .... **80,000 lbs.**
BALAST .... **30,000 lbs.**
TOTAL WEIGHT 110,000 lbs.

FIELD SPEED

Gear	Standard MPH	Optional MPH
1st	3.58	2.90
2nd	4.25	3.44
3rd	5.41	4.38
4th	6.39	5.17
5th	7.44	6.20
6th	8.81	7.13
Road	7 to 18	6 to 15

POWER TRAIN SPECIFICATIONS

ENGINE	Detroit Diesel 16V92T 760 Horsepower at 1900 RPM
TRANSMISSION	Twin Disc TD-61-2609 6 speed Power Shift
AXLES	Clark D-85840 Planetary wheel ends, 26 x 8 air brakes
TIRES	United 38 x 35 Dualed 96'' diameter, 40'' wide

The building of the tractor was a major logistical problem in itself explained Ron Harmon. "We had a work force of around 80 qualified staff and over 50 of them were involved with the construction and assembly of the 747 at different stages. The tractor was just physically too large to build on an assembly line and too cumbersome to move around so we built the whole tractor in one place. Parts were taken to the tractor and it was literally built from the ground upwards."

"All the chassis and frame work along with the cab and all iron work were fabricated at Havre. The engine, transmission, axles, electrical and hydraulic plumbing etc were bought in. We tried to use as many off the shelf components as possible to enable the end user, in case of a breakdown, to repair the machine and get going again with as little down time as possible."

The tractor rolled out of the factory on a snowy January day in 1978 and was moved down to California where it was showcased at the Tulare Farm Show, a major equipment show held in February 1978. It was then moved to the Rossi Brothers farm at Bakersfield, California where it was demonstrated to farmers and the world's press. It was here the 747 was hooked to a purpose built deep soil cultivator fitted with 15- legged rippers and astounded its sceptics by working 15 acres an hour.

The Rossi brothers had previously been using two Caterpillar D9s to deep cultivate their land and incorporate the cotton residue. It took these two machines at least five passes

CLOCKWISE FROM TOP:

> When the Flexi-Coil cultivator for the 747 was built it measured 100 foot wide which proved too big so it was cut down to 80 foot, yet in a working day it can keep two 60 foot seeders busy.

> Access to the massive Detroit Diesel 16V92T 16-cylinder engine in V configuration with dual turbochargers is relatively simple.

to condition the soil. The Cats were fitted with 5-legged rippers working around 15 acres in a 10 hour day. The 747 could work in one hour what these machines could work in a day. The 747 could travel at around 6 mph and with this higher speed the soil could be broken up more easily, which incorporated the waste residue better.

This tractor more than met the Rossi Brothers expectations, crop yields increased by as much as 10%, labor and machinery costs were reduced and subsequent crops could be planted in the best possible conditions.

Overall profitability increased as Ron explained. "It was planned for this machine to quickly pay for itself. The Rossi Brothers said it may not have paid for itself in the first year but it definitely had by the end of the second year. Together we had worked to produce a tractor that could go into full production at an affordable cost and prove its worth. Sadly we

CLOCKWISE FROM TOP:

> A mammoth task was moving the tractor. Two banks of sand were built either side of the lowboy and the tractor driven over the lowboy. The tractor was jacked up and the wheels removed, then the whole machine was lowered onto the deck.

> Once loaded up, the epic journey back home began. It took nearly 12 days to haul the tractor back from Florida to Havre in Montana.

> The tractor on the right is a Big Bud 600/50 rated at 600hp weighing in at 60,000 lbs – 27.6 ton is a big powerful machine but compared to the 116,480 lbs - 52 ton rated up to 900hp, the 600/50 is quite small.

never built another tractor this big again."

In the mid 1980s due to a change in farming practices this huge tractor came on the market and Ron Harmon found a new owner, Jim Satori of Willowbrook Farms in Florida, who needed such a machine for his vegetable and fruit farm. The farm had a high peat content in its soil and water had to be pumped out into drainage canals; a difficult farm to work correctly. The 747 had no problem deep cultivating this farm. Satori also purchased a Series 4 Big Bud 740 rated at 740 engine hp and for several years these two tractors worked side by side until the 747's retirement in 1997. Ron Harmon once again purchased the tractor and bought it back to Montana on two large artic trailers especially adapted for the job.

Once back in Havre, the new owners, the Williams Brothers, set to rebuilding and repainting the 747. Gone were the original colors of white with black hood and exhaust stacks, in came a gleaming white tractor with new decals and lots of chrome. Over the winter of 1997 – 1998, 20 years after the original build, the 747 underwent a new look and minor engineering changes, nothing major. Ready for the spring cultivation work, the 747 looked like a brand new tractor.

The Williams brothers are very pleased with this big machine, their work rates have increased and now they do not have to work late into the night to get the jobs done. Working 80 acres every hour the 747 can keep two 60-foot seed drills working all day. In a 10 hour day the brothers are happy if they have worked and seeded 700 to 800 acres. Not a bad day's work.

ABOVE:

> The 52 ton giant with its 80 foot Flexi-Coil field cultivator looks relatively small on the wheat belt around Big Sandy. Travelling at around 6 mph pulling the 80 foot field cultivator it is possible to work more than an acre every minute and up to 700 acres in a 10 hour day is easily achievable.

CHAPTER 12

Big Bud tractor colors

Many Big Bud tractors were custom built machines to suit individual requirements with such diverse operations from agriculture, to dirt moving to laying fibre optic cables alongside the highway; with this in mind several of the Big Bud tractors were painted in different colors and liveries.

The standard Big Bud tractor color was Imron White with red and black decals.

The most famous of all the custom painted Big Bud tractors were the Bafus Blue Big Buds run by Oregon farmer Robert Bafus.

The Bafus Blue Big Bud

Robert Bafus farmed over 3,500 acres of prime arable land in Adams, Oregon with the majority of his land down to spring wheat. In the late 1970s Robert and his wife Freda were looking for a more powerful new tractor for their operation, the crawlers they were using at that time had become slow and outdated, they wanted something with power, big wheels and 4WD allowing them to speed up their whole operation.

Finding there was nothing large enough to suit their system they travelled to Big Bud Tractors Incorporated, Montana where they had heard that some of the biggest tractors in the world were being built.

Salesman Leo Bitz met Robert and his wife and together they spent time out on farms looking at tractors comparing various makes both on and off paper. Together they then went and watched various Big Bud tractors working in the field. Impressed with what he had seen, Robert Bafus placed an order for a Series 3 Big Bud 360/30, however before he signed the papers he stipulated the tractor be painted in his colors.

Leo agreed to this and the deal was struck.

As all Big Buds were factory sprayed and with decals already made, the newly ordered Bafus Big Bud had to be hand painted, the numbers, lettering and hood color scheme was undertaken by a local sign writer.

In all Robert Bafus had five Big Bud tractors painted in his favourite color 'Aqua Blue'. The Big Buds proved their worth and became the favourite tractors on the Bafus farm. All the Bafus farm buildings and various agricultural and road vehicles also appeared in the same 'Aqua Blue'.

The house and barns looked very distinctive in the predominant light blue and white color scheme.

Bafus had three Series 3 360/30s, one Series 3 525/50 and

CLOCKWISE FROM TOP:

> Robert Bafus with his trusted Massey Harris combine in 1956. It was probably the fantastic skies and tinted clouds over Oregon that inspired Robert to choose the Aqua Blue color scheme for his farm that later become known as Bafus Blue.

> The whole Bafus farmstead was painted in Aqua Blue from the house to the outbuildings and even the stock sheds.

one Series 4 370 painted Imron white and Aqua Blue. The 360/30s were the most used tractors on the farm whilst the 525/50 left the farm with very few hours on the clock. The 1987 Big Bud 370 spent time at the farm but was hardly ever used as Robert's health deteriorated and he retired soon after its delivery. When traded in, the new owner at Hogeland, Montana had a quality tractor well maintained and in excellent condition with virtually no hours on the clock.

The Bafus Blue name and Bafus Blue Big Buds have become an important part of Montana's big tractor history.

CLOCKWISE FROM TOP:

> Doug and Dick Hamilton farm 7,000 acres in Montana on the Canadian border; with 50% summer fallow and 50% cropping growing spring wheat and malting barley. They farm with both Big Bud and Steiger tractors, preferring the two Bafus Blue Buds they currently farm with.

> The 370 serial number 87-303-370 had virtually no hours on the clock and had never been used on the land when the Hamiltons bought the tractor. The more modern Komatsu engine is far more fuel efficient than the older Cummins Series engines, the full Powershift transmission aids operator comfort as well as general tractor wear and tear.

> The Cummins powered Bafus Blue 360/30 is mainly used for cultivation operations, fitted with a Flexi Coil 41 foot chisel plow travelling at 6 –7 mph it can work in the region of 45 acres per hour.

The Iranian Big Buds

Three HN 350 Series 1 Big Buds were ordered in 1976 by a large farming corporation in Iran, the order specified they be painted in Chevrolet Yellow. The three tractors were completed during the winter months of early 1977 and exported to Iran where it is believed they are still working today.

Many rumours have circulated as to the whereabouts of these yellow Big Buds; are they still in existence or have they been scrapped. It was firmly believed that during the Iran Iraq War of 1980 - 1988 the tractors were commandeered by the Iranian military.

They were put to use hauling military hardware across the deserts and to recover military vehicles that had become bogged down in the soft sand.

For many years it had been virtually impossible to track down the three tractors or find their exact whereabouts due to the various conflicts in the country.

As far as anyone was aware they had not been exported or destroyed in the conflicts.

Reports have recently been received that one big yellow tractor fitting the description of a Series 1 Big Bud has been seen working in the Ahvaz Khorramshahr area of Iran, which is an area of irrigated alluvial plains off the Shat al Arrab water way, the delta of the Tigris and Euphrates, it has also been reported that a big yellow wheeled tractor of similar appearance has been seen operating in the plains to the west of Tehran near the Caspian Sea.

Unconfirmed, it does appear that two of the three Big Bud tractors are still working in agriculture and have survived the conflicts of recent years.

Using well recognised components readily available off the shelf it appears that these 39 year old machines have been kept running.

Iran is a big country and the farming community might locate the third machine very soon.

CLOCKWISE FROM TOP:

> Custom painting of Big Bud tractors was not too much of a problem and has become something of a cult following, which included the Iranian Big Buds.

> Three HN 350 Series I Big Buds were ordered in 1976 by a large farming corporation in Iran.

> The exact location of these tractors since the Iran-Iraq war of 1980 – 1988 is unknown but they are believed to be still working in Iran.

Yellow Big Buds

CLOCKWISE FROM TOP:

> In their day the Wagner WA-14 tractors were giants of the land. The high horsepower Big Bud tractors soon took the title of giants of the fields.

> The 650/50 pulling a 2520 Yielder No Till seed drill at around 7mph was operating on Joe Kinnie's farm in Colorado.

> Many Big Bud tractors were painted in custom liveries often for individual farmers' requests. With Big Bud tractor production being in low numbers it was quite easy to change a paint scheme without holding a production line up. This particular 650/50 is rated at 650 engine hp from a Detroit Diesel 12V92T dual turbocharged and aftercooled engine, and is fitted with a Twin Disc Powershift Transmission.

The Orange Big Buds

The Friggstad Big Bud

In 1980 the farm implement manufacturer Friggstad wanted to relocate a proportion of its cultivator manufacturing business to the USA from its hometown of Frontier in Saskatchewan, Canada. Friggstad could not source adequate labor in Frontier to supply a required work force of up to 100 people to run the facility, therefore the decision was made to look for new manufacturing premises in an area where a workforce was readily available.

Olaf Friggstad's son Dave moved a few miles south across the border into Montana where he set up a small facility next to Ron Harmon's tractor manufacturing plant in Havre.

At that particular time Ron Harmon was having trouble with Twin Disc, one of his major component suppliers, and Friggstad found out that for political reasons it could not build its cultivators outside Canada. For a short while the two joined forces designing and building scraper boxes for land levelling; these scrapers were painted orange and badged as Friggstad Scraper Pans.

Due to the non-delivery of transmission parts for Big Bud tractors, which caused considerable financial problems for Ron Harmon as he was not able to move his stock of nearly complete tractors, the banks were pressing him to sell off his inventory to release capital. Borg Warner, a large construction and finance company in Texas did not like the way the banks were treating the smaller companies that were starting to run into hard times, which was due partly to the poor agricultural economy of the early to mid 1980s. Borg Warner approached Ron Harmon and bailed Big Bud Incorporated out, purchasing much of the Big Bud inventory.

At the same time Borg Warner was trying to gain a contract with the US Federal Government, the Military were looking for 6,000 units to shift dirt at several major sites. Big Bud machines at that time were virtually unknown outside of the neighbouring states of Montana so Borg Warner purchased three scraper boxes and one Big Bud tractor all painted

CLOCKWISE FROM TOP:

> Friggstad's largest chisel plow was 80 foot; the Canadian company was a pioneer in field cultivators.

> A Friggstad scraper box built at Havre, Montana behind a Big Bud 525/50 fitted on industrial tires.

> A Big Bud 665 painted orange and when coupled to the Friggstad scraper boxes it was named a Friggstad Big Bud. The scraper box held 25 cubic yard of heaped dirt

orange. This industrial build unit was named Friggstad, which was put through its paces and demonstrated its capabilities moving dirt at a new military base.

The wheeled Friggstad tractor three-scraper pan combination moved four times the amount of dirt Caterpillar undertook with equivalent machines at the demonstration. Borg Warner and Big Bud were awarded the contract, but the next morning the contract was cancelled being awarded to the much larger Caterpillar Company.

Impressed with the Big Bud tractor work rates, Borg Warner took several more tractors to Houston, Texas, to move dirt on other major contracts. These tractors were built and classed as industrial machines called EMS Big Buds, (Earth Moving Systems). The tractors up to 665hp proved their worth working thousands of trouble free hours until they were retired and bought back to Havre to be refurbished during the late 1990s.

Borg Warner used both orange and white Big Bud tractors, the EMS tractors, the 525hp 525/84, 650hp 650/50, and the 665hp Big Bud 665 were all heavy duty industrial build tractors designed to push dirt, pull scraper boxes and able to pull cultivation equipment on all types of land from agricultural, mining and construction reclamation work.

CLOCKWISE FROM TOP:

> The Big Bud 665 in the orange livery powered by a turbocharged Detroit Diesel 12V92T rated at 665hp

> The cab is equipped with its own hand operated hydraulic jacking system for raising and lowering it in the field.

> Borg Warner used both orange and white Big Bud tractors, the EMS tractors, the 525hp 525/84, the 650hp 650/50 and the 665hp Big Bud 665 were heavy duty industrial build tractors designed to both push dirt and pull scraper boxes.

CHAPTER 13

Industrial Big Bud tractors

High powered 4WD articulating machines had been used in heavy construction work for many years, during the 1950s and 60s small 50 to 100hp 2WD tractors had been struggling on the land. The Wagner brothers and then the Steiger brothers had seen these big machines working on land levelling and road construction operations and they could see the benefits that would be gained by adapting and building such machines for agricultural use.

The Wagner brothers converted their industrial units to agricultural machines during the mid to late 1950s while the Steiger brothers built a pivot steer unit from scratch; brothers Douglass and Maurice Steiger built a 4WD articulating tractor from the ground upwards to suit their specific farming needs.

By the end of the 1950s the benefits of both high power and traction were now available to both the construction industry and farmers alike. Wagner continued building both industrial and agricultural machines while Steiger concentrated on the agricultural market with a very few machines offered in yellow livery to the construction industry.

The new breed of big 4WD tractors proved to be ideal machines for land levelling work pulling scraper boxes, hauling dirt trailers or deep ripping on land reclamation. These machines had the benefit of being a multi-purpose machine, whereas many of the purpose built machines could only undertake one job with another machine having to be bought in for a separate operation, this was more evident with metal tracked tracked tractors.

Big Bud tractors had been primarily designed as agricultural units even though several machines, due to their correct power to weight ratio, had been used in various construction applications. The first official Big Bud to be used in the construction industry was probably the Friggstad Big Bud, a one-off machine adapted for a specific task.

The Big Bud 665 was painted orange and the Friggstad name was used on the hood to match the scraper boxes built by Friggstad. Together the combined units were demonstrated on several big land levelling operations to show what they could achieve. The scraper boxes could easily be removed to demonstrate the machines' versatility on deep ripping and other operations.

After watching various demonstrations a large construction company from Texas, Borg Warner, purchased nine Big Bud tractors plus several Friggstad scraper boxes for its operations in Texas. These machines were painted in both the Friggstad

ABOVE:

> Big Bud tractors are used for many operations; here a 450/50 is being used for laser landing levelling for flood irrigation.

orange and the Imron white and black livery and called EMS Big Buds, EMS denoting Earth Moving Systems.

Several of the EMS Big Buds varied from that first orange Friggstad Big Bud which was powered by a Detroit Diesel 12V92T rated at 650hp from its V12 engine. Several of the EMS order were based on the 525/50 with new pumps providing 600hp from a Cummins KTA 1150-C diesel, whilst others were based on the Big Bud 665 powered by a Detroit Diesel 12V92TA engine rated at 665hp.

Due to the success of the EMS Big Buds, further machines

CLOCKWISE FROM TOP:

> Looking back from the Cruiser Cab showing excellent visibility.

> A 1985 Big Bud 525/50 pulling a Bron HS-II single seven-foot shank burying plastic piping. This type of outfit can be used for laying drainage pipe or water pipe across open land. During the dry period of the early years of 2000 in and around Montana big tractors were used to lay water pipe to supply drinking water to drought stricken pastures.

> A Bron HS-II fitted to a 665hp industrial Big Bud, this combination can sink water pipe or cabling with ease. To avoid frost damage, the water pipe had to be laid to a uniform depth of six foot.

were commissioned for construction operations. Apart from their power and great traction, these wheeled tractors had an added advantage over many machines of the day working in the dirt moving and construction industry.

One example was when tracked tractors were being used to pull single shank rippers along the sides of highways to bury telecommunication and fibre optic cables, but they had several major drawbacks.

As metal tracked machines cannot travel on the highway due to the damage they can cause to the highway's surface, when junctions had to be crossed, or the machine had to cross over the highway or travel from one job to another, wooden sleepers had to be laid down for the machines to cross, this often proved a timely and costly exercise.

When moving from one job to the next or just travelling a few miles down the highway, the tracked machine had to be loaded onto a low loader and moved, again a timely costly exercise. The wheeled Big Buds had none of these problems, they were quicker at work, they could travel on the highway with no need for added protection to the road surface and they could travel reasonable distances at speeds of approximately 20 mph with no problems.

The Big Bud tractor proved to be an ideal machine for this type of work and is still in use today burying cables across America or water pipes across fields to provide stock with water. Quite often when a Big Bud is exchanged in to the Big Equipment LLC, it is quickly and expertly refurbished and equipped for such work.

CLOCKWISE FROM TOP:

> Many Big Buds were fitted with heavy-duty industrial tires for construction work; the Big Bud was ideal in this industry for pulling 25 cubic yard scraper boxes.

> The new breed of big 4WD agricultural tractors proved to be ideal machines for land reclamation and levelling after mining operations, or creating lagoons and reservoirs pulling scraper boxes and hauling dirt trailers.

> In 1980 a new industrial tractor was released, the Big Bud 525/84 described as a tractor that singularly or in combination could move, push, scrape, pile, blade or dozer materials such as dirt, coal and other materials.

> A Big Bud 525/50 on dual wheels can work in places a tracked tractor would find the going difficult. The largest Krac plow, if conditions are right, can cut to a depth of 10 foot 6 inch weighing in at 20,000 lbs - nine ton. Krac recommend for a 525hp rubber tired tractor the KC750 which can cut to a depth of 7 foot 3 inch weighing 14,000 lb – 6.25 ton.

CLOCKWISE FROM TOP:

> Ron Harmon's new company, Big Equipment LLC founded in 1995, was established to sell, maintain and refurbish big 4WD tractors, in particular Big Bud tractors that he had been involved with since 1975. Owners who had worked Big Bud tractors for many years, including EMS, traded in their older machines to Big Equipment LLC, often purchasing a newer updated model. After refurbishing the traded in machine, the Big Bud tractor leaves the Big Equipment LLC as a virtually brand new machine.

> Big Equipment was approached with this particular order as the team could offer wheeled machines for the work with horsepower ratings and work rates others could not match. The 665hp heavy-duty tractors could plow a trench over five feet deep and were capable of carrying all the cable reels. A special rack had been designed on the front of the tractor to guide the fibre optic cable, which was fed into the trench behind the tractor.

BIG BUD

CLOCKWISE FROM TOP:

> In 1998 a big order came into Big Equipment from C R Fredrick Inc for three 665 tractors and one smaller machine, a 650/50 to dig trenches with two smaller 525/50s undertaking minor duties. C R Fredrick used the machines to open up trenches and lay fibre optic cable in a 125 mile long trench for AT&T in southern California.

> To this day, and for the foreseeable future, Big Bud tractors are often the choice for contractors who want high horsepower machines capable of a good day's work at an affordable price with minimal downtime to undertake demanding work.

> The Big Bud 665 machines were constructed from much heavier steel than previous models, the one and a half inch main frame plate guaranteed longevity and the ability to remove and replace the power train and all major components quickly, either in the workshop or out in the field, ensured that these tractors could be serviced and maintained easily whilst the tractors could be continually refurbished and upgraded.

> Two Big Bud 665 tractors plus two operators were used to do the trenching and laying of the cable. Previously four tracked machines had been used to do the same amount of work; the average speed of the tracked machine was half a mile per hour while the Big Bud could pull the plow at five foot deep at two mile per hour.

Notable dates

1961 – 1968 The Wagner Sales and Service depot in Havre, Montana was owned by Wilber Hensler. The main business of the company was to sell and service Wagner tractors, short lines and associated equipment throughout Northern Montana.. The workshop foreman at the depot at that time was Big Bud Nelson.

1968 New Years Eve 1968, John Deere purchased the production rights of the Wagner WA-14 and WA-17 tractors from the Wagner brothers in Portland, Oregon. The WA prefix standing for Wagner Agricultural.

1969 On the 1st January, both Hensler and Nelson were left pondering as to which way the company was to go now that they were left without a tractor to sell. Previously the company had been modifying various Wagner tractors, rubber mounting the cabs for increased operator comfort, equipping the tractors with dual wheels and many other modifications to bring the Wagner tractors up to date.

For a couple of years previously Bud Nelson had been toying with the idea of building a more powerful and heavier tractor than the earlier Wagner models and now he had an ideal opportunity to put his ideas into practice.

With rough plans drawn on pieces of paper and chalk drawings on the workshop floor, Hensler and Nelson designed their first tractor. By the end of 1969 the first Big Bud tractor was built which was powered by a Cummins engine. This first tractor was called the Big Bud HN-250 rated at 250 engine hp.

Hensler and Nelson continued to service and repair Wagner tractors whilst developing their new tractor and they changed the name of their business to the Northern Manufacturing Company.

1975 Early January, Ronald Harmon purchased the Northern Manufacturing Co. from Hensler and Nelson and he continued the manufacturing of Big Bud tractors. A new company evolved called Harmon's Manufacturing Company.

1978 Due to the growing popularity of Big Bud tractors across the Northern States of America the company name was changed to Big Bud Tractors Incorporated.

1979 Ronald Harmon purchased his father's truck stop and converted the buildings into a sales and service depot for Big Bud tractors. The Big Bud Sales and Service depot was a wholly owned subsidiary of Big Bud Inc.

1985 The Big Bud Sales and Service facility was closed down and the building became the new Case IH dealership owned by Meissner Brothers Inc. who had become equity partners with Big Bud Inc. Ron Harmon managed Big Bud affairs for the Meissner Brothers Inc.

1988 The Meissner Brothers Inc. took full ownership of the Big Bud dealership. At the same time two Indian Reservations, Fort Belnap and Rocky Boy under Government funding trained their members into management and general employment opportunities. A new company was formed owned equally by the Meissner Brothers Inc. and the Fort Belnap and Rocky Boy Indian reservations called Big Bud Manufacturing Incorporated.

1990 For various reasons the two Indian tribes pulled out of the deal and Meissner Brothers Inc. took full control of the company and Big Bud Manufacturing Incorporated ceased trading. Big Bud tractor production also ceased.

1995 Ron Harmon formed a new business the Big Equipment Company, to repair, rebuild and modify existing Big Bud tractors with a view to restarting the manufacture of Big Bud tractors.

2005 Williams Brothers begin the purchase of the Big Bud name and rights from Meissner Brothers Inc.

Big Bud time line

Big Bud tractor production dates

1969	The first Big Bud tractor the HN-250 - No 6901
1970	Big Bud Series 1 HN-320 tractors introduced - No 7000
1975	Big Bud KT Series introduced in Series 1 - No 7513
1977	Series 2 introduced which included both HN and KT Series tractors
1978	HN-360 tractor was the last in the HN Series - No 7878
1978	KT-525 tractor was the last in the KT Series No - 7879
1978	The 16V-747 built - No 7816
1979	Series 3 introduced, the 8V360 - No 7840
1986	Series 4 introduced, the 370 - No 86301
1988	The last Series 3 tractor a 650/84 - No 8803
1990	The last Series 4 tractor a 450 - No 9050

Older Big Bud tractors being refurbished for agricultural and industrial uses

Facility still in place to start building Series 5 Big Bud tractors

1970 – 1977

Big Bud Series 1 HN tractors

1975 – 1977

Big Bud Series 1 KT tractors

1977 – 1978

Big Bud Series 2 HN tractors

1977 – 1978

Big Bud Series 2 KT tractors

1977 / 1978

The Big Bud 16V-747 tractor

1979 – 1988

Big Bud Series 3 tractors

1986 – 1990

Big Bud Series 4 tractors

Company names and dates

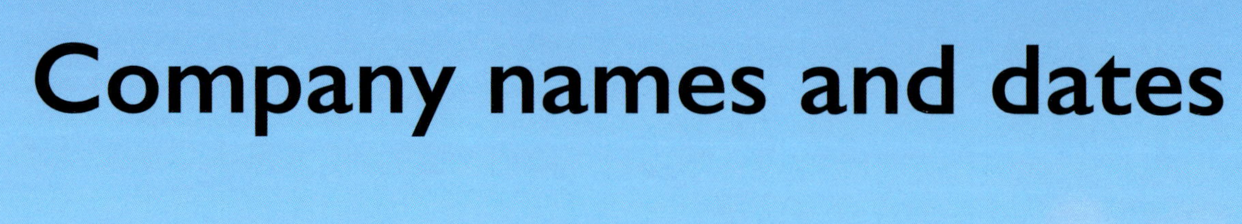

1961 – 1968	Wagner Sales and Service
1969 – 1974	The Northern Manufacturing Company
1975 – 1978	Harmon's Northern Manufacturing Company
1978 – 1985	Big Bud Tractors Incorporated
	Big Bud Sales and Service
1985 – 1988	Big Bud Tractors Incorporated
	Meissner Tractors Incorporated
1988 – 1990	Big Bud Manufacturing Incorporated
1990 –	Meissner Tractors Incorporated
1995 –	Big Equipment Company LLC

BIG BUD

Big Bud tractor specifications

Big Bud HN Series 1

Most HN Series 1 Big Buds were similar in overall appearance with only a few slight cosmetic changes; the simplest way to identify HN Series 1 models is viewed from the front with the straight up and down sides of the hood. The 'Cruiser Cab' features a full width glass to the operator's right and on his left where the door is situated there are three sections of glass.

Northern Manufacturing Co.

BIG BUD

ALL - WHEEL DRIVE TRACTOR

MODELS:
HN-250 HN-350 KT-450

BUILT BY

Northern Manufacturing Co.

HAVRE, MONTANA
Highway 2 West 406/265-6431

Big Bud Specifications

ENGINE:
- Cummins in-line diesel
- HN-250: NT855C280 — 280 G.H.P. @ 2100 RPM
- NT855P310 — 310 G.H.P. @ 2100 RPM
- HN-350: NTC 350 — 350 G.H.P. @ 2100 RPM

TRANSMISSION:
- Fuller RTO-12513 with Oil Coller
- Single lever shift
- 12 Speeds forward & 2 reverse

AXLES:
- HN-250: Clark 37,400 Planatary Drive
  23.222 : 1 Ration
- HN-350: Clark 70,500 Planatary Drive
  23.917 : 1 Ratio

CLUTCH:
- Spicer Angle Spring Type
- 15½-2 Plate with Transmission Brake

DRIVELINES:
- HN-250: Spicer 1700 Series x 8C
- HN-350: Spicer 1700 Series x 9C

TRANSFER CASE:
- Morse-Hyvo #824 Chain
- 6'' wide with 1'' pitch

HYDRAULICS:
- Webster Dual Pump driven directly off Crankshaft
- 20 gallons/minute for tractor @ 2100 RPM
- 20 gallons/minute for implement @ 2100 RPM
- Cross 3 spool self cancelling implement valve
- Gresen RFA-203 33 micron filter
- 42 U.S. Gallons tank capacity

STEERING:
- Charlyn Orbitrol System
- Articulating Frame
- 2-6'' Single acting cylinders with 4'' rams

TIRES:
- Singles or Duals
- 23.1 x 30
- 28.1 x 26
- 24.5 x 32 (HN-350 only)
- 66 x 43.00 - 25 Terra Tire (single only)

BRAKES:
- Disc type with double pads
- Air operated
- Driveline mounted

DRAWBAR:
- Swinging type on roller
- 2''x6'' steel with 1¾'' pin

FUEL CAPACITY:
- 525 U.S. Gallons

WEIGHT:
- HN-250: 34,000 lb.
- HN-350: 40,000 lb.

WHEEL BASE:
- HN-250: 141 inches
- HN-350: 144 inches

OVERALL LENGTH:
- HN-250: 259 inches
- HN-350: 264 inches

CAB:
- Rubber mounted
- Hydraulic tilt
- Built in air conditioning
- Tinted Glass

Big Bud KT Series 1

The Big Bud KT Series 1 are of similar design to the HN Series 1 models, from a distance without identifying marks it would be difficult to tell the difference. The HN tractors tended to be in the lower power range under 400hp whereas the KT tractors were over 400hp up to 525hp.

BIG BUD

Northern Manufacturing Co.

P.O. BOX 1111
HAVRE, MONTANA 59501

PH. [406] 265-6031
PH. [406] 265-6431

KT-450

KT-450 KT-525

ENGINE: Cummins in-line 6 cylinder diesel, Engine displacement - 1150 cubic inch, Horse power - 450, 525 at 2100 RPM.

KT 450

TRANSMISSION: Fuller Roadranger twin countershaft 13 speeds forward - 2 speeds reverse.

AXLES: Caterpillar planetary drive - 980 series.

CLUTCH: Spicer Angle Spring Button type 15½ inch twin disc with transmission brake.

DRIVELINES: Spicer - 1810 series x 9c Mechanics Universal Driveline joint.

TRANSFER CASE [DIVIDER BOX]: Power transfer box from transmission to driveline Hyvo 824 Morse power chain (1 in. pitch 6 in. wide).

WEIGHT: 43,000 pounds dry weight.

FUEL CAPACITY: 525 U.S. Gallons

HYDRAULICS: Dual pumps - 22 gal/min. to tractor steering, 22 gal/min. to implement control valve. Implement control valve is a 3-spool open center self-cancelling valve.

TIRES: Duals 30.5 x 32 Inside, 24.5 x 32 Outside.

HN-350

ENGINE: Cummins in-line 6 cylinder diesel, Engine displacement - 855 cubic inch, Horse power - 350 at 2100 RPM.

TRANSMISSION: Fuller Roadranger twin countershaft 13 speeds forward - 2 reverse.

AXLES: Caterpillar 980 series.

CLUTCH: Spicer Angle Spring Button type 15½ inch twin disc with transmission brake.

HN 350 RICE OPTION

HN 350 RICE OPTION

DRIVELINES: Spicer - 1710 series x 9c Mechanics Universal Driveline joint.

TRANSFER CASE [DIVIDER BOX]: Power transfer box from transmission to driveline Hyvo 824 Morse power chain (1 in. pitch 6 in. wide).

WEIGHT: 40,000 pounds dry weight.

FUEL CAPACITY: 525 U.S. Gallons

HYDRAULICS: Tandem pump - 20 gal/min. to tractor steering, 20 gal/min. to implement control valve. Implement control valve is a Cross 3-spool open center self-cancelling valve.

TIRES: 24.5 x 32 Duals.

HN-320

ENGINE: Cummins in-line 6 cylinder diesel, Engine displacement - 855 cubic inch, Horse power - 320 Horsepower at 2100 RPM.

TRANSMISSION: Fuller Roadranger twin countershaft 13 speeds forward - 2 speeds reverse.

AXLES: Caterpillar 528 series.

CLUTCH: Spicer Angle Spring button type 15½ inch twin disc with transmission brake.

DRIVELINES: Spicer - 1710 series x 7½c Mechanics Universal Driveline joint.

HYDRAULIC TILT CAB

TRANSFER CASE [DIVIDER BOX]: Power transfer box from transmission to driveline Hyvo 824 Morse power chain (1 in. pitch 6 in. wide).

WEIGHT: 36,000 pounds dry weight.

FUEL CAPACITY: 525 U.S. Gallons.

HYDRAULICS: Tandem pump - 20 gal/min. to tractor steering, 20 gal/min. to implement control valve. Implement control valve is a Cross 3-spool open center self-cancelling valve.

TIRES: 23.1 x 30 Duals.

REMOVABLE POWER SKID

Printed in U.S.A. by Griggs Printing & Publishing, Havre, Montana

Big Bud HN and KT Series 2

Series 2 HN and KT models are virtually identical; the decaling is the easiest way to identify between the two Series 2 tractors. The 'Cruiser Cab' is used once again, and updated newly styled tin work most noticeable from the front with the hood being angled inwards at the top and then straight down to the fenders for improved operator visibility. Once again the HN Series were under 400hp while the KT Series went from 400 to 525hp.

1977 BIG BUD

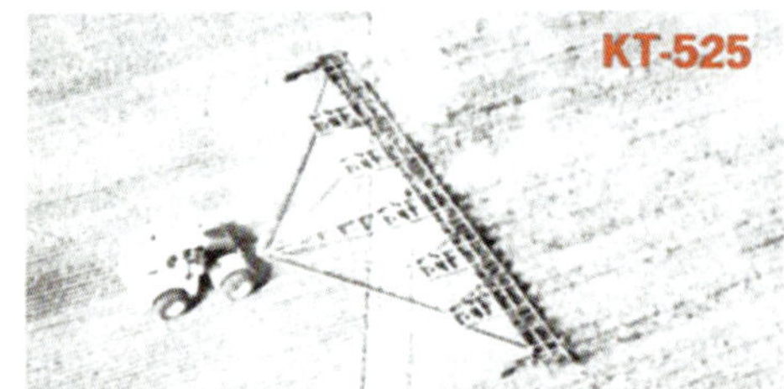

ENGINE:

TRANSMISSION:

TRANSFER CASE:
(DIVIDER BOX)

CLUTCH:

DRIVELINES:

AXLES:

STEERING:

HYDRAULICS:

WEIGHT:

FUEL CAPACITY:

ELECTRICAL:

Special Equipment Options:

Cruiser Cab® options:
Torsion suspension bucket-type seat. Air suspension bucket-type seat. CB radio. Refrigerator.

Northern Manufacturing Company
P.O. Box 1111 Havre, Montana USA 59501
Telephone (406) 265-5887

Cummins in-line 6-cylinder diesel, turbocharged and after-cooled (KT-450 turbocharged only). Displacement ... 1150 cu. in. (18.85 liters), Horsepower ... 450 & 525 at 2100 RPM

Fuller Roadranger RT 12513. Twin counter shaft. Forward speeds ... 13. Reverse speeds ... 2.
(see transmission option)

Big Bud designed and manufactured transmission-to-drive-line power box, moves with rear frame for reduced wear points.
Power chain ... Hyvo 824 Morse, 1 in. pitch, 6 in. width; on Morse sprockets.

Spicer angle-spring, self-adjusting, twin trapezoid-type discs. Size ... 15½ in. (39.74 cm). Super-duty, disc-type transmission brake.

Spicer drivelines with Spicer 1810 series and 9c Mechanics Universal driveline joints.

Heavy duty Caterpillar 980 series. Planetary drive with full floating ring and sun gears, three planets on caged roller bearings, Duo-Cone® seals.
(see axle option)

Turning radius ... 26 ft. (7.92 meters). Steering wheel revolutions ... 4½ lock to lock. Control ... 35 gal./min. orbital motor with 75-cu. in. displacement per steering wheel revolution.
(see steering option)

Dual pumps (2 separate units), 22 gal./min. to steering, 22 gal./min. to implement control valve, 4-spool open center, 35 gal./min. implement valve, 50 gal./min. hydraulic filter. Large volume lines for less noise and heat.
(see hydraulic options)

41,000 lb. (dry weight) 18,636 kg.,
58,500 lb. (maximum ballasted weight) 26,590 kg.

U.S. Gallons ... 550 Liters ... 2081.36

12 volt DC negative ground system, 24 volt DC starter system, 75 amp alternator

Cummins in-line 6-cylinder diesel, turbocharged and after-cooled. Displacement ... 855 cu. in. (14.01 liters), Horsepower ... 360 at 2100 RPM

Fuller Roadranger RT 12513. Twin counter shaft. Forward speeds ... 13. Reverse speeds ... 2.
(see transmission option)

Big Bud designed and manufactured transmission-to-drive-line power box, moves with rear frame for reduced wear points.
Power chain ... Hyvo 824 Morse, 1 in. pitch, 6 in. width; on Morse sprockets.

Spicer angle-spring, self-adjusting, twin trapezoid-type discs. Size ... 15½ in. (39.74 cm). Super-duty, disc-type transmission brake.

Spicer drivelines with Spicer 1810 series and 9c Mechanics Universal driveline joints.

Heavy duty Caterpillar 980 series. Planetary drive with full floating ring and sun gears, three planets on caged roller bearings, Duo-Cone® seals.
(see axle option)

Turning radius ... 26 ft. (7.92 meters). Steering wheel revolutions ... 4½ lock to lock. Control ... 35 gal./min. orbital motor with 75-cu. in. displacement per steering wheel revolution.
(see steering option)

Tandem pump (single unit), 24 gal./min. to steering, 24 gal./min. to implement control valve, 4-spool open center, 35 gal./min. implement valve, 50 gal./min. hydraulic filter. Large volume lines for less noise and heat.
(see hydraulic options)

40,000 lb. (dry weight) 18,181 kg.,
58,500 lb. (maximum ballasted weight) 26,590 kg.

U.S. Gallons ... 550 Liters ... 2081.36

12 volt DC negative ground system, 12 volt DC starter system, 75 amp alternator

Cummins in-line 6-cylinder diesel, turbocharged and after-cooled. Displacement ... 855 cu. in. (14.01 liters), Horsepower ... 320 at 2100 RPM

Fuller Roadranger RT 12513. Twin counter shaft. Forward speeds ... 13. Reverse speeds ... 2.
(see transmission option)

Big Bud designed and manufactured transmission-to-drive-line power box, moves with rear frame for reduced wear points.
Power chain ... Hyvo 824 Morse, 1 in. pitch, 6 in. width; on Morse sprockets.

Spicer angle-spring, self-adjusting, twin trapezoid-type discs. Size ... 15½ in. (39.74 cm). Super-duty, disc-type transmission brake.

Spicer drivelines with Spicer 1710 series and 8½c Mechanics Universal driveline joints.

Heavy-duty Caterpillar 528 series. Planetary drive with full floating ring and sun gears, three planets on caged roller bearings, Duo-Cone® seals.
(see axle option)

Turning radius ... 26 ft. (7.92 meters). Steering wheel revolutions ... 4½ lock to lock. Control ... 35 gal./min. orbital motor with 75-cu. in. displacement per steering wheel revolution.
(see steering option)

Tandem pump (single unit), 24 gal./min. to steering, 24 gal./min. to implement control valve, 4-spool open center, 35 gal./min. implement valve, 50 gal./min. hydraulic Large volume lines for less noise and heat.
(see hydraulic options)

36,000 lb. (dry weight) 16,363 kg.

U.S. Gallons ... 550 Liters ... 2081.36

12 volt DC negative ground system, 12 volt DC starter system, 75 amp alternator

Hydraulic package option:

50 gal./min. implement pump
60 gal./min. implement valve
80 gal./min. hydraulic filter

Transmission option:

Automatic transmission available. Details on request.

Tire options:

A wide variety of tires and wheel setups are available to meet varied field and crop conditions. Options include R1, R2 and steel cord and belted tires.

Steering option:

3 turns, lock to lock. Control ... 35 gal./min. orbital motor with 100 cu. in. displacement per steering wheel revolution.

Dress-up accessory options:

Chromed accessories available for exhaust and filter cover components. Details on request.

Axle option:

Caterpillar 980 series, torque proportioning.

KT-525

KT-450

	KT-525	KT-450
ENGINE:	Cummins in-line 6-cylinder diesel, turbocharged and aftercooled. Displacement ... 1150 cu. in. (18.85 liters). Horsepower ... 525 (392 kw) at 2100 RPM — 600 (448 kw) optional engine available. Donaldson aircleaner system with aspirator.	Cummins in-line 6-cylinder diesel, turbocharged. Displacement ... 1150 cu. in. (18.85 liters). Horsepower ... 450 (335 kw) at 2100 RPM. Donaldson aircleaner system with aspirator.
TRANSMISSION:	Fuller Roadranger RT 12513. Twin counter shaft. Forward speeds ... 13. Reverse speeds ... 2. (see transmission option)	Fuller Roadranger RT 12513. Twin counter shaft. Forward speeds ... 13. Reverse speeds ... 2. (see transmission option)
TRANSFER CASE: (DIVIDER BOX)	Big Bud designed and manufactured transmission-to-driveline power box, moves with rear frame for reduced wear points. Power chain ... Hyvo 824 Morse, 1 in. pitch, 6 in. width; on Morse sprockets.	Big Bud designed and manufactured transmission-to-driveline power box, moves with rear frame for reduced wear points. Power chain ... Hyvo 824 Morse, 1 in. pitch, 6 in. width; on Morse sprockets.
CLUTCH:	Spicer angle-spring, self-adjusting, twin trapezoid-type discs. Size ... $15\frac{1}{2}$ in. (39.74 cm). Super-duty, disc-type transmission brake.	Spicer angle-spring, self-adjusting, twin trapezoid-type discs. Size ... $15\frac{1}{2}$ in. (39.74 cm). Super-duty, disc-type transmission brake.
DRIVELINES:	Spicer drivelines with Spicer 1810 series and 9c Mechanics Universal driveline joints.	Spicer drivelines with Spicer 1810 series and 9c Mechanics Universal driveline joints.
AXLES:	Heavy-duty Caterpillar 980 series. Planetary drive with full floating ring and sun gears, three planets on caged roller bearings, Duo-Cone® seals.	Heavy-duty Caterpillar 980 series. Planetary drive with full floating ring and sun gears, three planets on caged roller bearings, Duo-Cone® seals.
STEERING:	Turning radius ... 26 ft. (7.92 meters). Steering wheel revolutions ... $4\frac{1}{2}$ lock to lock. Control ... 35 gal./min. orbital motor with 75-cu. in. displacement per steering wheel revolution.	Turning radius ... 26 ft. (7.92 meters). Steering wheel revolutions ... $4\frac{1}{2}$ lock to lock. Control ... 35 gal./min. orbital motor with 75-cu. in. displacement per steering wheel revolution.
HYDRAULICS:	Dual pumps (2 separate units) 32 gal./min. to steering. 50 gal./min. to implement control valve, 4-spool open center, implement valve, 100 gal./min. hydraulic filter. Large volume lines for less noise and heat.	Dual pumps (2 separate units) 32 gal./min. to steering. 50 gal./min. to implement control valve, 4-spool open center, implement valve, 100 gal./min. hydraulic filter. Large volume lines for less noise and heat.
WEIGHT:	41,000 lb. (dry weight) 18,636 kg. 58,500 lb. (maximum ballasted weight) 26,590 kg.	41,000 lb. (dry weight) 18,636 kg. 58,500 lb. (maximum ballasted weight) 26,590 kg.
FUEL CAPACITY:	U.S. Gallons ... 550. Liters ... 2081.36.	U.S. Gallons ... 550. Liters ... 2081.36.
ELECTRICAL:	12 volt DC negative ground system, 24 volt DC starter system, 75 amp alternator.	12 volt DC negative ground system, 24 volt DC starter system, 75 amp alternator.
TIRES:	30.5 x 32 duals, R-1.	30.5 x 32 inside, 24.5 x 32 outside, R-1.

HN-360

HN-320

	HN-360	HN-320
ENGINE:	Cummins in-line 6-cylinder diesel, turbocharged and aftercooled. Displacement ... 855 cu. in. (14.01 liters). Horsepower ... 360 (269 kw) at 2100 RPM. Donaldson aircleaner system with aspirator.	Cummins in-line 6-cylinder diesel, turbocharged and aftercooled. Displacement ... 855 cu. in. (14.01 liters). Horsepower ... 320 (239 kw) at 2100 RPM. Donaldson aircleaner system with aspirator.
TRANSMISSION:	Fuller Roadranger RT 12513. Twin counter shaft. Forward speeds ... 13. Reverse speeds ... 2. (see transmission option)	Fuller Roadranger RT 12513. Twin counter shaft. Forward speeds ... 13. Reverse speeds ... 2. (see transmission option)
TRANSFER CASE: (DIVIDER BOX)	Big Bud designed and manufactured transmission-to-driveline power box, moves with rear frame for reduced wear points. Power chain ... Hyvo 824 Morse, 1 in. pitch, 6 in. width; on Morse sprockets.	Big Bud designed and manufactured transmission-to-driveline power box, moves with rear frame for reduced wear points. Power chain ... Hyvo 824 Morse, 1 in. pitch, 6 in. width; on Morse sprockets.
CLUTCH:	Spicer angle-spring, self-adjusting, twin trapezoid-type discs. Size ... $15\frac{1}{2}$ in. (39.74 cm). Super-duty, disc-type transmission brake.	Spicer angle-spring, self-adjusting, twin trapezoid-type discs. Size ... $15\frac{1}{2}$ in. (39.74 cm). Super-duty, disc-type transmission brake.
DRIVELINES:	Spicer drivelines with Spicer 1720 series and 9c Mechanics Universal driveline joints.	Spicer drivelines with Spicer 1710 series and $8\frac{1}{2}$c Mechanics Universal driveline joints.
AXLES:	Heavy-duty Caterpillar 980 series. Planetary drive with full floating ring and sun gears, three planets on caged roller bearings, Duo-Cone® seals.	Heavy-duty Caterpillar 528 series. Planetary drive with full floating ring and sun gears, three planets on caged roller bearings. Duo-Cone® seals.
STEERING:	Turning radius ... 26 ft. (7.92 meters). Steering wheel revolutions ... $4\frac{1}{2}$ lock to lock. Control ... 35 gal./min. orbital motor with 75-cu. in. displacement per steering wheel revolution.	Turning radius ... 26 ft. (7.92 meters). Steering wheel revolutions ... $4\frac{1}{2}$ lock to lock. Control ... 35 gal./min. orbital motor with 75-cu. in. displacement per steering wheel revolution.
HYDRAULICS:	Tandem pump (single unit) 32 gal./min. to steering. 32 gal./min. to implement control valve, 4-spool open center, implement valve, 70 gal./min. hydraulic filter. Large volume lines for less noise and heat.	Tandem pump (single unit) 32 gal./min. to steering. 32 gal./min. to implement control valve, 4-spool open center, implement valve, 70 gal./min. hydraulic filter. Large volume lines for less noise and heat.
WEIGHT:	39,000 lb. (dry weight) 17,727 kg. 48,000 lb. (maximum ballasted weight) 21,818 kg.	36,000 lb. (dry weight) 16,363 kg.
FUEL CAPACITY:	U.S. Gallons ... 550. Liters ... 2081.36.	U.S. Gallons ... 550. Liters ... 2081.36.
ELECTRICAL:	12 volt DC negative ground system, 12 volt DC starter system, 75 amp alternator.	12 volt DC negative ground system, 12 volt DC starter system, 75 amp alternator.
TIRES:	24.5 x 32 duals, R-1.	23.1 x 30 duals, R-1.

KT-400

Cummins in-line 6-cylinder diesel, turbocharged. Displacement ... 1150 cu. in. (18.85 liters). Horsepower ... 450 (335 kw) at 2100 RPM. Donaldson aircleaner system with aspirator.

Fuller Roadranger RT 12513. Twin counter shaft. Forward speeds ... 13. Reverse speeds ... 2. (see transmission option)

Big Bud designed and manufactured transmission-to-driveline power box, moves with rear frame for reduced wear points. Power chain ... Hyvo 824 Morse, 1 in. pitch, 6 in. width; on Morse sprockets.

Spicer angle-spring, self-adjusting, twin trapezoid-type discs. Size ... 15½ in. (39.74 cm). Super-duty, disc-type transmission brake.

Spicer drivelines with Spicer 1720 series and 9c Mechanics Universal driveline joints.

Heavy-duty Caterpillar 980 series. Planetary drive with full floating ring and sun gears, three planets on caged roller bearings, Duo-Cone® seals.

Turning radius ... 26 ft. (7.92 meters). Steering wheel revolutions ... 4½ lock to lock. Control ... 35 gal./min. orbital motor with 75-cu. in. displacement per steering wheel revolution.

Dual pumps (2 separate units) 32 gal./min. to steering. 50 gal./min. to implement control valve, 4-spool open center, implement valve, 100 gal./min. hydraulic filter. Large volume lines for less noise and heat.

40,000 lb. (dry weight) 18,181 kg. 50,000 lb. (maximum ballasted weight) 22,727 kg.

U.S. Gallons ... 550. Liters ... 2081.36.

12 volt DC negative ground system, 24 volt DC starter system, 75 amp alternator.

24.5 x 32 duals, R-1.

Optional equipment available:

Cruiser Cab options:
Torsion suspension bucket-type seat. Air suspension bucket-type seat. CB radio. Refrigerator. Scannex rear view TV scanner. Buddy seat and tool box.

Transmission option:
Twin Disc power shift automatic (clutch-free) transmission.

Tire options:
A wide variety of tires and wheel arrangements are available to handle a variety of field and crop conditions. Options include R-1, R-2 and steel cord and belted tires.

Dress-up accessory options:
Chromed accessories available for exhaust and filter cover components.

The Big Bud 16V-747

The biggest agricultural tractor in the world, the Big Bud 16V-747 in its 1998 livery of the Williams Brothers. The 747 is instantly recognisable by its sheer size, big chrome stacks and new graphics give the tractor a more modern look.

Big Bud 16V-747 specifications

ENGINE

DETROIT DIESEL 16V92T

Maximum rated brake horsepower	760 (567 KW) at 1900 RPM
Aspiration	Dual turbochargers
Type	16 cylinders, V-configuration, 2 cycle
Displacement	1472 cu. in. (24.14 liters)
Torque	2372 lb. ft. (3216 N.m) at 1400 RPM

ENGINE WARRANTY

Detroit Diesel Allison Warranty	36-month or 2500-engine-hour warranty (whichever first occurs). No charge for parts. No charge for labor.

TORQUE CONVERTER

TWIN DISC 8FLW-1801

Type	18 in. (45.72 cm.) circuit with lock-up clutch
Blading	MS-380

TRANSMISSION

TWIN DISC TD-61-2609, full power shift

Forward speeds	6
Reverse speeds	1

AXLES

CLARK D-85840

Type	Limited-slip differential with planetary wheel ends

BRAKES

26 in. x 8 in. (66.04 cm. x 20.32 cm.) airbrakes, all wheels

HYDRAULICS

System type	Load-sensing system allowing maximum pump flow to steering and implement valve.
Maximum flow	65 gal./min.(245.96 liters/min.)
Tank capacity	150 gallons (567.64 liters)
Relief pressure	2000 psi
Implement valve	4-spool open center, 60 gal./min. (227.05 liters/min.)
Steering	Orbital motor with 100 cu. in. (1.64 liters) displacement per steering wheel revolution
Steering cylinders	Double action dual cylinders, 4.5 in. (11.43 cm.) bore

ELECTRICAL

Starter system	24 volt
All other electrical	12 volt
Alternator	75 amp

CRUISER™ CAB

Unitized construction, pressurized with clear-view design. Standard features: Air conditioning, heater, windshield wipers (front and rear), safety tinted glass, deluxe vinyl-cloth adjustable swivel bucket-type seat, AM/FM 8-track stereo system, 40-channel CB radio, decelerator pedal, adjustable hydraulic controls, perforated vinyl foam insulation and padded acoustical floor mats, buddy seat and tool box.

WEIGHT

95,000 lb. (38,636 kg.) estimated shipping weight

130,000 lb. (49,895 kg.) ballasted weight

FUEL CAPACITY

850 U.S. gallons (3216.65 liters)

DIMENSIONS

Wheelbase	16 feet, 3 in. (4.95 meters)
Height (top of cab)	14 feet (4.27 meters)
Length (frame)	27 feet (8.23 meters)
Length (end of drawbar)	28 feet, 6 in. (8.69 meters)
Width (over fenders)	13 feet, 4 in. (4.06 meters)
Width (over duals)	20 feet, 10 in. (6.35 meters)

TIRES

UNITED 35 x 38 duals, 95.8 in. (243.3 cm.) diameter x 39.6 in. (100.58 cm.) width.

OPTIONS

Scannex rear view closed-circuit TV scanner

Tractor performance monitor

Refrigerator

Big Bud Series 3 Models

Big Bud Series 3 high horsepower models

Out of all the Big Bud tractors built in Havre, the Series 3 525/50 was the most popular and it was this model that gained the name 'The Big Bud Five and a Quarter.

Looking at the larger models, the styling looked the same as the smaller Series 3 tractors, the main visual difference being the larger frame size to give added weight so extra horsepower could be utilized efficiently.

650/50

PRODUCT DESCRIPTION

ENGINE
DETROIT DIESEL 12V92T diesel

Maximum rated brake horsepower	650 (500 kW) 2100 RPM
Aspiration	Dual turbocharged and aftercooled
Type	12 cylinders, V configuration, 2 cycle
Bore and Stroke	4.84 in. x 5 in. (123 mm. x 127 mm.)
Displacement	1104 cu. in. (18.1 liters)
Torque at peak	1915 ft./lb. (2461 N.m) at 1300 RPM
Injectors	Cam operated, unit type, clean tip design
Compression ratio	17 to 1
Lubrication	Force feed to all bearings and gears
Detroit Diesel Engine Warranty	36-month or 2500-engine-hour warranty, (whichever first occurs). No charge for parts. No charge for labor.

CHASSIS
BIG BUD-designed and manufactured chassis featuring Terra-Torque™ chassis equalization system for oscillation and weight distribution.

Frame members	1.5 in. (3.85 cm.) thick steel
Articulation	Single-point
Equalization	Eight-point Terra-Torque™ design
Drawbar	Center-point with 5-point sideswing lock

TRANSMISSION
Twin Disc (TD-92-2610) power shift, lock-up gear type, direct transmission (clutchless). Incorporates transmission and power divider coupled to a Twin Disc torque converter.

Forward speeds	9
Reverse speeds	2

TORQUE CONVERTER

Twin Disc 8FLW-1801 type	18 in. (45.72 cm.) circuit with lock-up clutch
Blading	MS-380

AXLES
Clark D-75830, standard differential with planetary wheel ends.

BRAKES
) x 7 in. (66.04 cm. x 20.32 cm.) airbrakes, front axle; and 25 in. (9.84 cm.) diameter rear drive line disc brake.

HYDRAULICS
Closed Center Two-pump System
Steering Circuit—Pressure compensated using 0-36 gals./min. (0-136.27 liters/min.) axial piston pump, 35 gals./min. (132.49 liters/min.) orbital motor and two 4-in. (10.16 cm.) diameter double-acting cylinders.

Implement Circuit—Pressure and flow compensated using 0-36 gals./min. (0-136.27 liters/min.) axial piston pump and 40 gals./min. (151.41 liters/min.) implement with four metering spools.
Filtration—100 gals./min. (378.54 liters/min.) capacity at 10 micron
Tank Capacity—100 gals. (378.54 liters)

ELECTRICAL

Starter system	24 volt, negative ground
All other electrical	12 volt, negative ground
Alternator	75 amp

CRUISER™ CAB
Unitized construction, pressurized with clear-view design. Standard features: Air conditioner, heater, windshield wipers (front and rear), safety tinted glass, deluxe vinyl-cloth adjustable swivel bucket-type seat, AM/FM 8-track stereo system, decelerator pedal, perforated vinyl foam insulation and padded acoustical floor mats, buddy seat and tool box.

INSTRUMENT PANEL
Tachometer, engine pyrometer, oil pressure, water temperature, engine oil temperature, Sentinel pre-shutdown engine alarm, transmission oil temperature, front differential temperature, rear differential temperature, power divider temperature, converter temperature and transmission pressure, voltmeter, brake warning light, air pressure gauge, dome light, turn and emergency flasher lights, cigarette lighter.

WEIGHT
60,000 lb. (38,636 kg.) estimated shipping weight
74,000 lb. (49,895 kg.) maximum ballasted weight

FUEL CAPACITY
700 U.S. gallons (3216.65 liters)

DIMENSIONS

Wheel Base	13 ft. 4 in.	(4.06 meters)
Height (top of cab)	12 ft. 10 in.	(3.91 meters)
Length (end of drawbar)	22 ft. 10 in.	(6.96 meters)
Length (overall)	22 ft.	(6.71 meters)
Width (over fenders)	10 ft. 10 in.	(3.30 meters)
Width (over duals)	17 ft. 7 in.	(5.36 meters)
Clearance (lowest spot)	20.5 in.	(52.07 cm.)
Turning radius	30 ft.	(9.14 meters)

TIRES

Duals .... 35.5 x 32, 79.2 in. (201.1 cm.) diameter x 35 in. (88.9 cm.) width

650/50 FIELD SPEED DATA

With Twin Disc TD-92-2610 power shift (clutch-free) transmission, Clark D-75830 axles, 35.5 x 32 dual tires:

Transmission Gear & Ratio		Field Speeds (miles per hour)
	Field Range	
1st	4.30:1	4.23
2nd	3.63:1	5.01
3rd	2.85:1	6.38
4th	2.41:1	7.55
5th	2.07:1	8.79
6th	1.75:1	10.39
	Transport Range	
7th	1.66:1	10.92
8th	1.11:1	16.46
9th	.80:1	22.65

OPTIONS

Dickey-john monitoring system
Scannex rear-view TV scanner
Refrigerator
Chrome muffler and filter cover components

Big Bud Tractors, Inc.
P.O. Box 1111
Havre, Montana USA 59501
Phone: (406) 265-5457

525/50 525/20

PRODUCT DESCRIPTION

ENGINE
CUMMINS KTA 1150-C diesel

Maximum rated brake horsepower	525 (392 kw) at 2100 RPM
Aspiration	Turbocharged and aftercooled
Type	6 cylinder, in-line 4 cycle
Displacement	1150 cu. in. (18.85 liters)
Bearings	Seven 5½ in. (14 cm.) diameter main bearings with 4 in. (10.2 cm.) diameter connecting rod bearings
Cylinder block	Alloy cast iron with replaceable wet-type cylinder liners
Cylinder heads	Six individual cylinder heads with corrosion-resistant inserts on intake and exhaust valve seats. Drilled fuel supply and return lines
Lubrication	Force feed to all bearings and gears
Turbocharger	Scroll diffuser, side-mounted

CHASSIS
BIG BUD-designed and manufactured chassis featuring Terra-torque™ chassis equalization system for oscillation and weight distribution.

Articulation	Single-point
Equalization	Eight point Terra-torque™ design
Drawbar	Center-point with sideswing dampener

TRANSMISSIONS (Twin Disc or Fuller Roadranger available)
TWIN DISC power shift, lockup gear type, direct transmission (clutchless). Incorporates transmission and power divider coupled to a Twin Disc torque converter.

Forward speeds	6
Reverse speeds	1

FULLER Roadranger RT 12513, twin counter shaft

Forward speeds	12
Reverse speeds	2

CLUTCH (only with Fuller transmission)
SPICER super heavy-duty, 3600 lb. angle-spring, self-adjusting twin trapezoid-type discs.

Size	15½ in. (39.74 cm.) diameter

DRIVELINES
SPICER 1810 series (yoke type) and 9c Mechanics

DIVIDER BOX (with Fuller transmission only)
BIG BUD designed and manufactured transmission-to-driveline power box.

Ratio	1.625:1
Drive chain and sprockets	Hyvo 824 Morse drive chain, 1 in. (2.54 cm.) pitch, 6 in. (15.24 cm.) width on Morse sprockets
Articulation pins	Two 2½ in. (6.35 cm.) diameter pins on spherical ball bearings

AXLES
Clark D-75830 planetary drive axles. (6:1 planet ratio)
Clark D-75720 planetary drive axles. (4:1 planet ratio)

BRAKES
Wheel drum type. S-cam air-actuated.

STEERING

Steering wheel revolutions	6.5 turns, lock to lock

HYDRAULICS

System type	Load-sensing system allowing maximum pump flow to steering and implement valve.
Maximum flow	38 gals./min. (144 liters/min.)
Implement control valve	4-spool open center, 60 gals./min. (226.8 liters/min.) capacity, 2000 psi recommended, 3000 psi maximum
Steering	35 gals./min. (132.3 liters/min.) orbital motor with 50 cu. in. (.82 liters) displacement per steering wheel revolution
Filter capacity	100 gals./min. (378 liters/min.)

ELECTRICAL

Starter system	24 volt, negative ground
All other electrical	12 volt, negative ground
Alternator	75 amp

CRUISER™ CAB
Unitized construction, pressurized with clear view design. Standard features: Air conditioner, heater, windshield wipers (front and rear), safety tinted glass, tilt and telescoping steering wheel, adjustable swivel bucket-type seat, decelerator pedal, AM/FM 8-track stereo system, adjustable hydraulic controls, perforated vinyl foam insulation and padded acoustical floor mats.

INSTRUMENT PANEL
Tachometer, engine pyrometer, oil pressure, water temperature, engine oil temperature, transmission oil temperature, front differential temperature, rear differential temperature, power divider temperature, converter temperature and transmission pressure voltmeter, brake warning light, air pressure gauge, dome light, turn and emergency flasher lights, cigarette lighter.

WEIGHT

45,000 lb. (20,430 kg.) approximate dry weight
60,000 lb. (27,180 kg.) approximate maximum ballasted weight

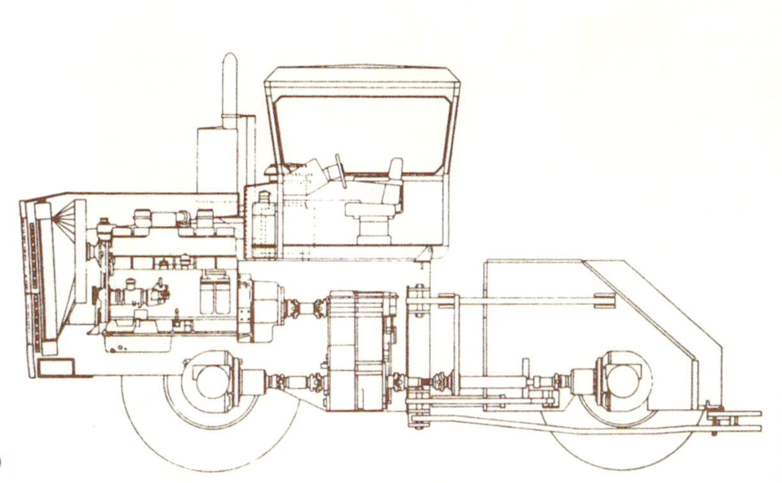

FUEL CAPACITY

550 U.S. gallons (2081.36 liters)

TIRES

Duals	30.5 x 32, R-1

DIMENSIONS

Wheel Base	150.0 in. (381 cm.)
Length	267.6 in. (675.6 cm.)
Height	156.0 in. (396.3 cm.)
Height (top of cab)	144.5 in. (367 cm.)
Width	120.0 in. (305.5 cm.)
Clearance (lowest spot)	14.5 in. (36.8 cm.)
Turning Radius	26 ft. (7.92 meters)

525/20 FIELD SPEED DATA

With Fuller Roadranger RT 12513 transmission, Clark 75720 axles, 30.5 x 32 dual tires:

Transmission Gear & Ratio	Field Speeds (miles per hour)
1st —— 8.35:1	1.59
2nd —— 6.12:1	2.17
3rd —— 4.56:1	2.91
4th —— 3.33:1	3.92
5th UD- 2.86:1	4.63
5th DR- 2.47:1	5.37
6th UD- 2.10:1	6.32
6th DR- 1.81:1	7.33
7th UD- 1.56:1	8.50
7th DR- 1.35:1	9.82
8th UD- 1.16:1	11.43
8th DR- 1.00:1	13.26

525/50 FIELD SPEED DATA

With Twin-Disc TD-61 2610 power shift (clutch-free) transmission, Clark 75830 axles, 30.5 x 32 dual tires:

Transmission Gear & Ratio	Field Speeds (miles per hour)
1st —— 4.30:1	3.78
2nd —— 3.63:1	4.48
3rd —— 2.85:1	5.71
4th —— 2.41:1	6.75
5th —— 2.07:1	7.86
6th —— 1.75:1	9.29

SUGGESTED RETAIL BASE PRICE

525/50 Twin Disc Power Shift	$143,000
525/20 Fuller Roadranger	$126,000

OPTIONS

Cruiser Cab options:

CB radio	$ 250
Refrigerator	300
Scannex rear view TV scanner	1,700
Buddy seat and tool box	120

Dress-up accessory options:

Chrome muffler and filter cover components	$ 250

TIRES

30.5 x 32, R-2, duals	$ 2,500

450/50 450/20

PRODUCT DESCRIPTION

ENGINE
CUMMINS KT 1150-C diesel

Maximum rated brake horsepower	450 (335 kw) at 2100 RPM
Aspiration	Turbocharged
Type	6 cylinder, in-line, 4 cycle
Displacement	1150 cu. in. (18.85 liters)
Bearings	Seven 5½ in. (14 cm.) diameter main bearings with 4 in. (10.2 cm.) diameter connecting rod bearings
Cylinder block	Alloy cast iron with replaceable wet-type cylinder liners
Cylinder heads	Six individual cylinder heads with corrosion-resistant inserts on intake and exhaust valve seats. Drilled fuel supply and return lines
Lubrication	Force feed to all bearings and gears
Turbocharger	Scroll diffuser, side-mounted

CHASSIS
BIG BUD designed and manufactured chassis featuring Terra-Torque™ chassis equalization system for oscillation and weight distribution.

Articulation	Single-point
Equalization	Eight-point Terra-Torque™ design
Drawbar	Center-point with sideswing dampener

TRANSMISSIONS (Twin Disc or Fuller Roadranger available)
TWIN DISC power shift, lockup gear type, direct transmission (clutchless). Incorporates transmission and power divider coupled to a Twin Disc torque converter.

Forward speeds	6
Reverse speeds	1

FULLER Roadranger RT 12513, twin counter shaft

Forward speeds	12
Reverse speeds	2

CLUTCH (only with Fuller transmission)
SPICER super heavy-duty, 3600 lb. angle-spring, self-adjusting twin trapezoid-type discs.

Size	15½ in. (39.74 cm.) diameter

DRIVELINES
SPICER 1810 series (yoke type) and 9c Mechanics

DIVIDER BOX (with Fuller transmission only)
BIG BUD designed and manufactured transmission-to-driveline power box.

Ratio	1.625:1
Drive chain and sprockets	Hyvo 824 Morse drive chain, 1 in. (2.54 cm.) pitch, 6 in. (15.24 cm.) width on Morse sprockets
Articulation pins	Two 2½ in. (6.35 cm.) diameter pins on spherical ball bearings

AXLES
Clark D-75720 planetary drive axles (4:1 planet ratio)

BRAKES
Wheel drum type. S-cam air-actuated.

STEERING

Steering wheel revolutions	6.5 turns, lock to lock

HYDRAULICS

System type	Load-sensing system allowing maximum pump flow to steering and implement valve.
Maximum flow	38 gals./min. (144 liters/min.)
Implement control valve	4-spool open center, 60 gals./min. (226.8 liters/min.) capacity, 2000 psi recommended, 3000 psi maximum
Steering	35 gals./min. (132.3 liters/min.) orbital motor with 50 cu. in. (.82 liters) displacement per steering wheel revolution
Filter capacity	100 gals./min. (378 liters/min.)

ELECTRICAL

Starter system	24 volt, negative ground
All other electrical	12 volt, negative ground
Alternator	75 amp

CRUISER™ CAB
Unitized construction, pressurized with clear view design. Standard features: Air conditioner, heater, windshield wipers (front and rear), safety tinted glass, tilt and telescoping steering wheel, adjustable swivel bucket-type seat, decelerator pedal, AM/FM 8-track stereo system, adjustable hydraulic controls, perforated vinyl foam insulation and padded acoustical floor mats.

INSTRUMENT PANEL
Tachometer, engine pyrometer, oil pressure, water temperature, engine oil temperature, transmission oil temperature, front differential temperature, rear differential temperature, power divider temperature, converter temperature and transmission pressure, voltmeter, brake warning light, air pressure gauge, dome light, turn and emergency flasher lights, cigarette lighter.

WEIGHT

42,000 lb. (19,026 kg.) approximate dry weight
50,000 lb. (22,650 kg.) approximate maximum ballasted weight

FUEL CAPACITY

550 U.S. gallons (2081.36 liters)

TIRES

Inside duals	30.5 x 32, R-1
Outside duals	24.5 x 32, R-1

DIMENSIONS

Wheel Base	150.0 in. (381 cm.)
Length	267.6 in. (675.6 cm.)
Height	156.0 in. (396.3 cm.)
Height (top of cab)	144.5 in. (367 cm.)
Width	120.0 in. (305.5 cm.)
Clearance (lowest spot)	14.5 in. (36.8 cm.)
Turning Radius	26 ft. (7.92 meters)

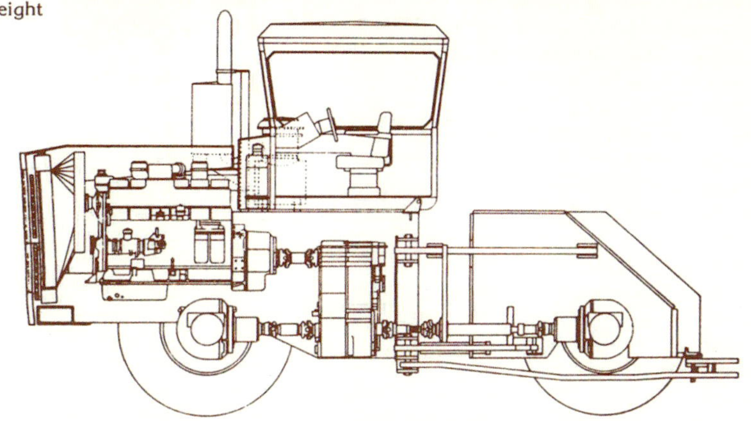

450/20 FIELD SPEED DATA

With Fuller Roadranger RT 12513 transmission, Clark 75720 axles, 24.5 x 32 dual tires:

Transmission Gear & Ratio	Field Speeds (miles per hour)
1st —8.35:1	1.59
2nd —6.12:1	2.17
3rd —4.56:1	2.91
4th —3.33:1	3.92
5th UD—2.86:1	4.63
5th DR—2.47:1	5.37
6th UD—2.10:1	6.32
6th DR—1.81:1	7.33
7th UD—1.56:1	8.50
7th DR—1.35:1	9.82
8th UD—1.16:1	11.43
8th DR—1.00:1	13.26

450/50 FIELD SPEED DATA

With Twin Disc TD-61-2610 power shift (clutch-free) transmission, Clark 75720 axles, 30.5 x 32 & 24.5 x 32 dual tires:

Transmission Gear & Ratio	Field Speeds (miles per hour
1st —4.30:1	3.67
2nd—3.63:1	4.35
3rd —2.85:1	5.54
4th —2.41:1	6.55
5th —2.07:1	7.63
6th —1.75:1	9.02

SUGGESTED RETAIL BASE PRICE

450/50 Twin Disc Power Shift	$132,000
450/20 Fuller Roadranger	$118,000

OPTIONS

Cruiser Cab options:

CB radio	$ 250
Refrigerator	300
Scannex rear view TV scanner	1,700
Buddy seat and tool box	120

Dress-up accessory options:

Chrome muffler and filter cover components	$ 250

TIRES

30.5 x 32 inside duals, R-2 and 24.5 x 32 outside duals, R-2	$ 1,900
30.5 x 32 duals, R-1	1,500
30.5 x 32 duals, R-2	3,500

PRODUCT DESCRIPTION

400/30 400/20

ENGINE

Detroit Diesel 8V-921 diesel

Maximum rated brake horsepower	400 (299 kw) at 2100 RPM
Aspiration	Turbocharged
Type	V-8, 2 cycle
Bore and Stroke	4.84 in. x 5 in. (123 mm. x 127 mm.)
Displacement	736 cu. in. (12.07 liters)
Injectors	Cam operated, unit type, clean tip design
Compression ratio	19 to 1
Lubrication	Force feed to all bearings and gears
Turbocharger	Vaneless, front top center mounted

CHASSIS

BIG BUD designed and manufactured chassis featuring Terra-Torque™ chassis equalization system for oscillation and weight distribution.

Articulation	Single-point
Equalization	Eight-point Terra-Torque™ design
Drawbar	Center-point with sideswing dampener

TRANSMISSIONS (Twin Disc or Fuller Roadranger available)

TWIN DISC power shift, lockup gear type, direct transmission (clutchless). Incorporates transmission and power divider coupled to a Twin Disc torque converter.

Forward speeds	6
Reverse speeds	1

FULLER Roadranger RT 12513, twin counter shaft

Forward speeds	12
Reverse speeds	2

CLUTCH (only with Fuller transmission)

SPICER super heavy-duty, 3600 lb. angle-spring, self-adjusting twin trapezoid-type discs.

Size	15½ in. (39.74 cm.) diameter

DRIVELINES

SPICER 1810 series (yoke type) and 9c Mechanics

DIVIDER BOX (with Fuller transmission only)

BIG BUD designed and manufactured transmission-to-driveline power box.

Ratio	1.625:1
Drive chain and sprockets	Hyvo 824 Morse drive chain, 1 in. (2.54 cm.) pitch, 6 in. (15.24 cm.) width on Morse sprockets
Articulation pins	Two 2½ in. (6.35 cm.) diameter pins on spherical ball bearings

AXLES

Clark D-75720 planetary drive axles (4:1 planet ratio)

BRAKES

Wheel drum type. S-cam air-actuated.

STEERING

Steering wheel revolutions	6.5 turns, lock to lock

HYDRAULICS

System type	Load-sensing system allowing maximum pump flow to steering and implement valve.
Maximum flow	38 gals./min. (144 liters/min.)
Implement control valve	4-spool open center, 60 gals./min. (226.8 liters/min.) capacity, 2000 psi recommended, 3000 psi maximum
Steering	35 gals./min. (132.3 liters/min.) orbital motor with 50 cu. in. (.82 liters) displacement per steering wheel revolution
Filter capacity	100 gals./min. (378 liters/min.)

ELECTRICAL

Starter system	24 volt, negative ground
All other electrical	12 volt, negative ground
Alternator	75 amp

CRUISER™ CAB

Unitized construction, pressurized with clear view design. Standard features: Air conditioner, heater, windshield wipers (front and rear), safety tinted glass, tilt and telescoping steering wheel, adjustable swivel bucket-type seat, decelerator pedal, AM/FM 8-track stereo system, adjustable hydraulic controls, perforated vinyl foam insulation and padded acoustical floor mats.

INSTRUMENT PANEL

Tachometer, engine pyrometer, oil pressure, water temperature, engine oil temperature, transmission oil temperature, front differential temperature, rear differential temperature, power divider temperature, converter temperature and transmission pressure, voltmeter, brake warning light, air pressure gauge, dome light, turn and emergency flasher lights, cigarette lighter.

WEIGHT

40,000 lb. (18,181 kg.) dry weight
50,000 lb. (22,727 kg.) maximum ballasted weight)

FUEL CAPACITY

550 U.S. gallons (2081.36 liters)

TIRES

Duals	24.5 x 32, R-1

DIMENSIONS

Wheel Base	150.0 in. (381 cm.)
Length	267.6 in. (675.6 cm.)
Height	156.0 in. (396.3 cm.)
Height (top of cab)	144.5 in. (367 cm.)
Width	120.0 in. (305.5 cm.)
Clearance (lowest spot)	14.5 in. (36.8 cm.)
Turning Radius	26 ft. (7.92 meters)

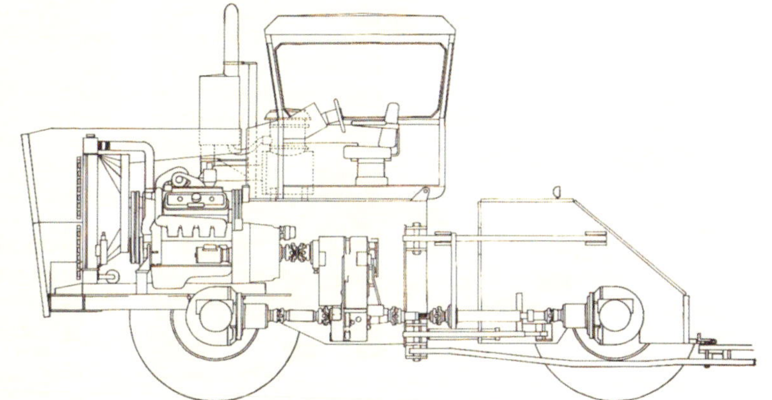

400/20 FIELD SPEED DATA

With Fuller Roadranger RT 12513 transmission, Clark 75720 axles, 24.5 x 32 dual tires:

Transmission Gear & Ratio	Field Speeds (miles per hour)
1st —8.35:1	1.59
2nd —6.12:1	2.17
3rd —4.56:1	2.91
4th —3.33:1	3.92
5th UD—2.86:1	4.63
5th DR—2.47:1	5.37
6th UD—2.10:1	6.32
6th DR—1.81:1	7.33
7th UD—1.56:1	8.50
7th DR—1.35:1	9.82
8th UD—1.16:1	11.43
8th DR—1.00:1	13.26

400/30 FIELD SPEED DATA

With Twin Disc TD-61-1154 power shift (clutch-free) transmission, Clark 75720 axles, 24.5 x 32 dual tires:

Transmission Gear & Ratio	Field Speeds (miles per hour)
1st —5.27:1	3.97
2nd—4.46:1	4.69
3rd —3.67:1	5.61
4th —3.09:1	6.77
5th —2.62:1	7.98
6th —2.15:1	9.72

SUGGESTED RETAIL BASE PRICE

400/30 Twin Disc Power Shift	$117,000
400/20 Fuller Roadranger	$109,000

OPTIONS

Cruiser Cab options:

CB radio	$ 250
Refrigerator	300
Scannex rear view TV scanner	1,700
Buddy seat and tool box	120

Dress-up accessory options:

Chrome muffler and filter cover components	$ 250

TIRES

24.5 x 32 duals, R-2	$ 1,500
30.5 x 32 duals, R-1	2,600

360/30 360/10

PRODUCT DESCRIPTION

ENGINE

CUMMINS NTA 885-C-360 diesel

Maximum rated brake horsepower	360 (269 kw) at 2100 RPM
Aspiration	Turbocharged and aftercooled
Type	6 cylinder, in-line, 4 cycle
Displacement	855 cu. in. (14 liters)
Bearings	Seven 4½ in. (11.4 cm.) diameter main bearings with 3-1/8 in. (7.9 cm.) diameter connecting rod bearings
Cylinder block	Alloy cast iron with replaceable wet-type cylinder liners
Cylinder heads	Three separate heads with corrosion-resistant inserts on intake and exhaust valve seats. Drilled fuel supply and return lines
Lubrication	Force feed to all bearings and gears
Turbocharger	Scroll diffuser, side-mounted

CHASSIS

BIG BUD designed and manufactured chassis featuring Terra-Torque™ chassis equalization system for oscillation and weight distribution.

Articulation	Single-point
Equalization	Eight-point Terra-Torque™ design
Drawbar	Center-point with sideswing dampener

TRANSMISSIONS (Twin Disc or Fuller Roadranger available)

TWIN DISC power shift, lockup gear type, direct transmission (clutchless). Incorporates transmission and power divider coupled to a Twin Disc torque converter.

Forward speeds	6
Reverse speeds	1

FULLER Roadranger RT 12513, twin counter shaft

Forward speeds	12
Reverse speeds	2

CLUTCH (only with Fuller transmission)

SPICER super heavy-duty, 3600 lb. angle-spring, self-adjusting twin trapezoid-type discs.

Size	15½ in. (39.74 cm.) diameter

DRIVELINES

SPICER 1810 series (yoke type) and 9c Mechanics

DIVIDER BOX (with Fuller transmission only)

BIG BUD designed and manufactured transmission-to-driveline power box.

Ratio	1.625:1
Drive chain and sprockets	Hyvo 824 Morse drive chain, 1 in. (2.54 cm.) pitch, 6 in. (15.24 cm.) width on Morse sprockets
Articulation pins	Two 2½ in. (6.35 cm.) diameter pins on spherical ball bearings

AXLES

Clark D-75720 planetary drive axles (4:1 planet ratio)

BRAKES

Wheel drum type. S-cam air-actuated.

STEERING

Steering wheel revolutions	6.5 turns, lock to lock

HYDRAULICS

System type	Load-sensing system allowing maximum pump flow to steering and implement valve.
Maximum flow	38 gals./min. (144 liters/min.)
Implement control valve	4-spool open center, 60 gals./min. (226.8 liters/min.) capacity, 2000 psi recommended, 3000 psi maximum
Steering	35 gals./min. (132.3 liters/min.) orbital motor with 50 cu. in. (.82 liters) displacement per steering wheel revolution
Filter capacity	100 gals./min. (378 liters/min.)

ELECTRICAL

Starter system	24 volt, negative ground
All other electrical	12 volt, negative ground
Alternator	75 amp

CRUISER™ CAB

Unitized construction, pressurized with clear view design. Standard features: Air conditioner, heater, windshield wipers (front and rear), safety tinted glass, tilt and telescoping steering wheel, adjustable swivel bucket-type seat, decelerator pedal, AM/FM 8-track stereo system, adjustable hydraulic controls, perforated vinyl foam insulation and padded acoustical floor mats.

INSTRUMENT PANEL

Tachometer, engine pyrometer, oil pressure, water temperature, engine oil temperature, transmission oil temperature, front differential temperature, rear differential temperature, power divider temperature, converter temperature and transmission pressure, voltmeter, brake warning light, air pressure gauge, dome light, turn and emergency flasher lights, cigarette lighter.

WEIGHT

39,000 lb. (17,727 kg.) dry weight
48,000 lb. (21,818 kg.) maximum ballasted weight

FUEL CAPACITY

550 U.S. gallons (2081.36 liters)

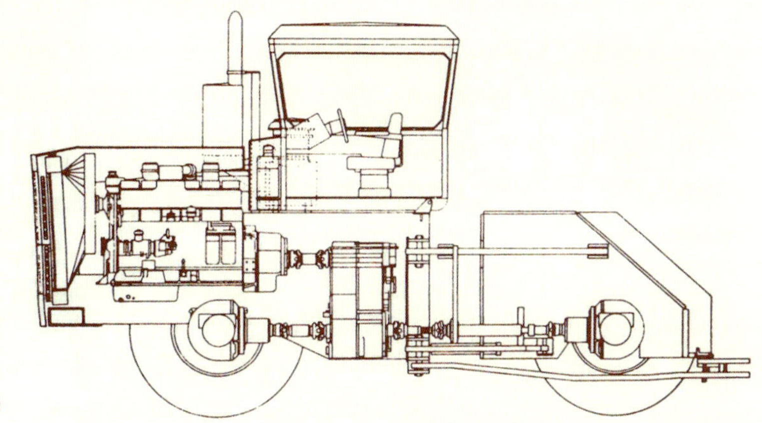

TIRES

Duals	24.5 x 32 duals, R-1

DIMENSIONS

Wheel Base	150.0 in. (381 cm.)
Length	267.6 in. (675.6 cm.)
Height	156.0 in. (396.3 cm.)
Height (top of cab)	144.5 in. (367 cm.)
Width	120.0 in. (305.5 cm.)
Clearance (lowest spot)	14.5 in. (36.8 cm.)
Turning Radius	26 ft. (7.92 meters)

360/10 FIELD SPEED DATA

With Fuller Roadranger RT 12513 transmission, Clark 75720 axles, 24.5 x 32 dual tires:

Transmission Gear & Ratio	Field Speeds (miles per hour)
1st —8.35:1	1.59
2nd —6.12:1	2.17
3rd —4.56:1	2.91
4th —3.33:1	3.92
5th UD—2.86:1	4.63
5th DR—2.47:1	5.37
6th UD—2.10:1	6.32
6th DR—1.81:1	7.33
7th UD—1.56:1	8.50
7th DR—1.35:1	9.82
8th UD—1.16:1	11.43
8th DR—1.00:1	13.26

360/30 FIELD SPEED DATA

With Twin Disc TD-61-1154 power shift (clutch-free) transmission, Clark 75720 axles, 24.5 x 32 dual tires:

Transmission Gear & Ratio	Field Speeds (miles per hour)
1st —5.27:1	3.97
2nd—4.46:1	4.69
3rd —3.67:1	5.70
4th —3.09:1	6.77
5th —2.62:1	7.98
6th —2.15:1	9.72

SUGGESTED RETAIL BASE PRICE

360/30 Twin Disc Power Shift	$111,000
360/10 Fuller Roadranger	$103,000

OPTIONS

Cruiser Cab options:

CB radio	$ 250
Refrigerator	300
Scannex rear view TV scanner	1,700
Buddy seat and tool box	120

Dress-up accessory options:

Chrome muffler and filter cover components	$ 250

TIRES

24.5 x 32 duals, R-2	$ 1,500

The Bigger Big Buds

There was never a sales promotion leaflet produced for the 600hp models that were designated 600/50, basically this particular model was the 525 main frame and cab with the engine opened up to 600hp. The 525/84 was a heavy-duty tractor for the construction industry. The 650/50 had a larger main frame, again to maximise the power to weight ratio. The Big Bud 665 was exactly the same as the Big Bud 650/50 except that the 665 had a hard nose grille that had ½ inch steel that caused the tractor to weigh an extra 1,000 pounds more than the 650/50 and was used in the construction industry.

525/84

In 1980 a new industrial tractor was released, the Big Bud 525/84 was described as a tractor that singularly or in combination could move, push, scrape, pile, blade or dozer materials such as dirt, coal and other raw materials with ease.

Big Bud 665

The mighty 665hp Big Bud 655 powered by a Detroit Diesel 12V92TA engine.

400/30
BIG BUD

EARTHMOVING SYSTEMS

SCRAPER TRAIN.

665-2025 TRAIN. TRAIN.

EMS

665-2025 TRACTOR SCRAPER

ENGINE
- Detroit 12V92TA Diesel
- 665 Horsepower (436 KW) @ 2100 RPM
- Turbocharged, 2 Cycle, 12 Cylinder, V Configuration for High Performance, Durability and Fuel Efficiency

POWER TRAIN
- Dependable, Powershift Transmission with Hydraulic Shifter for Gear, Down & Upshift Inhibiter & Converter Lock-Up
- Rigid Axles with Limited Slip Differential for Constant 4 Wheel Drive Traction
- Reliable Wheel End Planetaries

SERVICEABILITY
- Serviceability is Built-In with Hinged Cab & Hood that Provide Easy Access to Power Train Components
- Individually Removeable Universal Components to Reduce Maintenance Time

OPERATOR COMFORT & SAFETY
- Rubber Isolated ROPS/FOPS Cab is Fully Enclosed & Sound Suppressed
- Mechanical & Air Adjustable Seat with 180° Rotation Provides Maximum Comfort
- Full Instrumentation
- Tilt & Telescoping Steering Wheel

CAPACITY
- 20.0 cu. yd. Struck SAE
- 25.0 cy. yd. Heaped SAE

PRODUCTIVITY
- Low Profile Self-Loading with Double Acting Cylinders
- Low Angle Full Width Cutting Edge
- Low Maintenance Resulting From Simple Design
- Scraper Can Be Coupled Together Behind Single Power Unit for More Efficient, Lower Cost Operation

Bud 525/84
Preliminary Specifications

ENGINE

CUMMINS KTA 1150-C 525 diesel

*Maximum horsepower	525 at 2100 RPM
**Flywheel horsepower	473 at 2100 RPM
Governed RPM	2100
Maximum torque	1575 ft. lbs. at 1500 RPM
Bore and stroke	6.25 in. x 6.25 in.
Number of cylinders	6
Displacement	1150 cu. in.

*Maximum horsepower based on SAE standard 1816, barometric pressure of 29.38 in. hg. (74.62 cm. hg.) and ambient temperature of 85°F (29.4°C) at maximum engine speed with fuel pump, water pump and lubricating oil pump.

**Net usable horsepower at engine flywheel under SAE standard J816 with standard engine fan, alternator, water pump, lubricating oil pump, air compressor and air cleaner.

TRANSMISSION

CLARK Model 8821 powershift. Full power shifting through hydraulically actuated multiple disc clutches.

Forward speeds	8
Reverse speeds	4

TORQUE CONVERTER

CLARK Model CL 16902 high efficiency single stage type; 2.85:1 torque multiplication ratio with lock-up clutch.

AXLES

CLARK Model D-75830 all-wheel drive with 6:1 planetary reduction in each wheel. Limited slip front and rear differentials. Cradle mounted rear axle — oscillation 15°

TRAVEL SPEEDS

	1st	2nd	3rd	4th	5th	6th	7th	8th
Forward, MPH	3.0	4.0	5.4	7.2	9.5	12.7	17.3	22.7
Reverse, MPH	3.0	5.4	9.5	17.3				

HYDRAULIC SYSTEM

PUMP: Axial piston, variable displacement, pressure compensated, torque converter mounted.

VALVE: Two-spool type with built-in adjustable relief valve.

Pump capacity at governed RPM	60 GPM
System relief pressure	18.50 PSI
System capacity	190 US gals.
Reservoir capacity	170 US gals.
Lift cylinders, double acting, two	6.00 in.
Tilt cylinders, double acting, one	8 in.

STEERING

BUD articulated frame/chassis. Pump is piston-type, torque converter mounted.

Pump capacity at governed RPM	60 GPM
System relief pressure	1800 PSI
Steer cylinders, double acting, two	5.25 in.
Articulation	Dual-point

TURNING RADIUS

Outside corner of blade	25 ft. 3 in.
Outside rear tire (curb to curb)	25 ft.
Angle of steer	80° total, lock-to-lock

BLADE

Cutting edges — hardened alloyed steel.

Width (over cutting edge)	16 ft. 8 in.
Height	5 ft. 1 in.
Max. clearance under cutting edge	3 ft. 7 in.
Max. depth cutting edge below ground	1 ft. 3 in.
Max. blade tilt	8 in.
Manual pitch adjustment	
Blade speed	raise 55 FPM 5.0 lower 41 FPM 6.8

Paint — Black Imron

BRAKES

SERVICE: Four-wheel straight air, shoe type

Diameter and width	20 in. x 7 in.

EMERGENCY: Four-wheel maxi-chamber. Automatically actuated by low air pressure or manually applied through dash-mounted control; audible and visual alarm.

WHEELS AND TIRES

Standard tires	Duals 30.5 x 32 LS2
Optional tires	Duals 35.5 x 32 LS2
Rim diameter and width	32 in. x 27 in.

STANDARD EQUIPMENT

Dual ladders & hand rails, reverse warning alarm, ROPS cab, seat belt, air ride bucket seat, AM-FM tape, air conditioner, tilt & telescoping steering wheel, rear hitch, ether starting aid, horn, hi-intensity lighting (4 front-4 rear), heavy duty alternator 110 amp.

GAUGES: Engine temperature, oil pressure, hourmeter, voltmeter, torque converter temperature, air pressure with low air pressure warning light and buzzer.

FILTERS: Engine oil, fuel, air (dry-type), hydraulic oil (suction & return) torque converter/transmission.

Paint — Energy Orange Imron

ELECTRICAL
12 volt, negative ground system.
Alternator .... 110 amp

FUEL CAPACITY
450 U.S. gallons

WEIGHT (approximate)
67,000 lb. dry weight
77,000 lb. working with hydroinflation
72,000 lb. working with counterweight*
*In lieu of hydroinflation.

SCRAPER SPECIFICATIONS

DIMENSIONS
Self-contained hydraulic control system
Tandem, 20 yd. struck
25 yd. heaped
Tail ejection
15 ft. 6 in. cutting blade
Overall Width .... 16 feet
Transport Width .... 12 feet
Paint — Energy Orange Imron
Weight 32.500 lbs. empty, each unit

OPTIONS
Goose neck hitch
Laser mounting provision

Big Bud Series 4

The last Big Bud tractors to be built were the Series 4 or Generation 4 as they were sometimes referred to.

The Series 4 Big Buds took on a completely new look which was designed to show that Big Bud tractors were back on the market with their high tech engines, electronic transmissions and were serious contenders in the modern, competitive market place for big 4WD articulated tractors.

Most of the Series 4 tractors were built to order, there were two basic main frames utilized to take engines from various manufacturers ranging from 360hp through to 740hp.

BIG BUD

DIMENSIONS AND DATA

WHEEL BASE . . . . 155"
LENGTH . . . . 291"
WIDTH . . . . 173" (24.5 x 32 Duals)
188" (30.5 x 32 Duals)
HEIGHT . . . . 146" (Top of Cab)
TURNING RADIUS . . . . 224" (Center Line of Tractor)
WEIGHT . . . . 5/8" Frame - 43,000
1" Frame - 45,000
60/40 Front/Rear Ratio
FUEL . . . . 550 U.S. Gals. Distributed equally in both sides of front frame.
ARTICULATION . . . . 40° Degrees
OSCILLATION . . . . 15° Degrees

CAB
Electric/Hydraulic tilt. Rubber mounted for sound and vibration dampening.

Precleaned and filtered air used for pressurization, eliminating dust and stale air.

Heated and air conditioned.

A fully adjustable deluxe seat is mounted on a Big Bud designed heavy-duty, locking swivel air-ride base.

AM/FM cassette stereo, tilt-telescopic steering wheel column.

INSTRUMENTS AND GAUGES
Engine tachometer, ground speed (mechanical), transmission gear indicator, engine oil pressure, engine water temperature, engine oil temperature, pyrometer, transmission oil temperature, transmission oil pressure, front and rear differential temperature, volt meter, air pressure, hour meter and clock.

WARNING LIGHTS
Fuel level, low voltage, engine oil pressure, transmission oil pressure, charge pump hydraulic oil pressure, engine coolant temperature, park brake, transmission oil temperature, engine air filter, hydraulic oil temperature, transmission oil filter, high beam indicator, low air pressure and sentinal.

SPEED
Ground Speed - 2100 RPM

Gear	Speed	Gear	Speed
1	2.6 mph	7	8.4 mph
2	3.2 mph	8	10.0 mph
3	3.9 mph	9	12.4 mph
4	4.8 mph	10	15.2 mph
5	5.7 mph	11	18.2 mph
6	7.0 mph	12	22.4 mph
1st R	3.4 mph	2nd R	6.2 mph

AXLES
Clark D - 70500 outboard planetary
Total reduction 19.133:1
LTD Slip (Optional)

TWO TIRE SIZES
30.5 x 32 and 24.5 x 32; 20.8 x 42 Triples; Options Available.

BRAKES
Air/Hydraulic (Front Axle Only)
Equipped with Micro-Lock

TRANSMISSION
Full power shift, 12 speeds forward and 2 reverse. Electric/hydraulic controlled. A manually operated wet master clutch is used for vehicle "inching" in 1st gear, forward or reverse.

HYDRAULIC SYSTEM
Closed center, pressure compensated open loop system. 4 remotes are standard. Lever type (ISO) couplers. Pump capacity 45 GPM at 21 RPM with 2500 PSI maximum operating pressure. 135 gallon reservoir in rear frame.

ELECTRICAL SYSTEM
Starter System - 24 volt neg. ground. All other electrical - 12 volt neg. ground. Alternator - 130 amp; Battery - 2 each 8D heavy duty.

SERVICEABILITY
The push button tilt cab and hood opens up the complete power train for easy access and service. The center section has only one point of

SPECIFICATIONS subject to change without notification.

CATERPILLAR 3406 B DITA (INDUSTRIAL ENGINE) Maximum Rate BHP **402 HP**
Maximum Torque @1200 RPM-1335 lb. ft.
Torque Rise 33.5%
TYPE . . . . 6 In Line, 4 Cycle
Water Cooled
Direct Injection
ASPIRATION . . . . Turbo Charged, After Cooled
BORE AND STROKE . . . . 5.4" x 6.5" - 137 x 165 mm
DISPLACEMENT . . . . 893 Cu. In. - 14.6 Liter
COMPRESSION RATIO . . . . 14.5 to 1

CUMMINS NTA-855-A . . . . Maximum Rated BHP **450 HP**
Maximum Torque @1400 RPM-1400 lb. ft.
Torque Rise 20%
TYPE . . . . 6 In Line, 4 Cycle
Water Cooled
Direct Injection
ASPIRATION . . . . Turbo Charged, After Cooled
BORE AND STROKE . . . . 5.5" x 6" — 140 x 152 mm
DISPLACEMENT . . . . 855 Cu. In. — 14 Liter
COMPRESSION RATIO . . . . 14 to 1

CUMMINS KTA 19P . . . . Maximum Rated BHP **450-700 HP**
Maximum Torque @1300 RPM-1650 lb. ft.
Torque Rise 20%
TYPE . . . . 6 In Line, 4 Cycle
Water Cooled
Direct Injection
ASPIRATION . . . . Turbo Charged, After Cooled
DISPLACEMENT . . . . 1150 Cu. In.

GRIGGS PRINTING & PUBLISHING, HAVRE, MT 59501

KOMATSU SA6D-125-1:	Maximum rated BHP Maximum torque Torque rise	370 H P 1012 FT/LBS. 15%
TYPE	6 in line, 4 cycle, water-cooled, direct injection.	
ASPIRATION	Turbo charged and after cooled	
BORE AND STROKE	4.92" x 5.91" — 125x150 mm	
DISPLACEMENT	674 cu. in.	
COMPRESSION RATIO	14.5 : 1	

CUMMINS NTA-855-A	Maximum Rated BHP Maximum Torque @ 1400 RPM Torque Rise	375 HP 1266 LB. FT. 35%
TYPE	6 In Line, 4 Cycle Water Cooled Direct Injection	
ASPIRATION	Turbo Charged and After Cooled	
BORE AND STROKE	5.5" x 6.0" — 140 x 152 mm	
DISPLACEMENT	855 Cu. In. — 14 Liter	
COMPRESSION RATIO	14 to 1	

CATERPILLAR 3406 B DITA (INDUSTRIAL ENGINE)	Maximum Rated BHP Maximum Torque @ 1200 RPM Torque Rise	360 HP 1150 LB. FT. 28%
TYPE	6 In Line, 4 Cycle Water Cooled Direct Injection	
ASPIRATION	Turbo Charged, After Cooled	
BORE AND STROKE	5.4" x 6.5" - 137 x 165 mm	
DISPLACEMENT	893 Cu. In. - 14.6 Liter	
COMPRESSION	14.5 to 1	

CUMMINS NTA-855-A	Maximum Rated BHP Maximum Torque @ 1400 RPM Torque Rise	400 HP 1200 LB. FT. 20%
TYPE	6 In Line, 4 Cycle Water Cooled Direct Injection	
ASPIRATION	Turbo Charged, After Cooled	
BORE AND STROKE	5.5" x 6" — 140 x 152 mm	
DISPLACEMENT	855 Cu. In. — 14 Liter	
COMPRESSION RATIO	14 to 1	

KOMATSU SA6D—140-1	Maximum rated BHP Maximum torque Torque rise	440 HP 1374 FT/LBS. 25%
TYPE	6 in line, 4 cycle, water-cooled, direct injection	
ASPIRATION	Turbo charged and after cooled	
BORE AND STROKE	5.51" x 6.50" — 140 x 165 m.m.	
DISPLACEMENT	930 cu. in.	15.2 liter
COMPRESSION RATIO	14 : 1	

CATERPILLAR 3406 B DITA (INDUSTRIAL ENGINE)	Maximum Rate BHP Maximum Torque @ 1200 RPM Torque Rise	402 HP 1335 LB. FT. 33.5%
TYPE	6 In Line, 4 Cycle Water Cooled Direct Injection	
ASPIRATION	Turbo Charged, After Cooled	
BORE AND STROKE	5.4" x 6.5" - 137 x 165 mm	
DISPLACEMENT	893 Cu. In. - 14.6 Liter	
COMPRESSION RATIO	14.5 to 1	

DETROIT 8V-92TA	Maximum Rated BHP Maximum Torque @ 1300 RPM Torque Rise	425 HP 1175 LB. FT. 10%
TYPE	V-8, 2-Cycle Water Cooled Direct Injected	
ASPIRATION	Turbo Charged Bypass Blower After Cooled	
BORE AND STROKE	4.84" x 5.0" - 122.9 x 127 mm	
DISPLACEMENT	736 Cu. In. - 12.1 Liter	
COMPRESSION RATIO	17 to 1	

KOMATSU SA6D170-A-1	Maximum Rated B.H.P. Maximum Torque @ 1400 RPM Torque Rise	739 HP 2234 LB. FT. 15%
TYPE	6 In Line, 4 Cycle Water Cooled Direct Injected	
ASPIRATION	Turbo Charged, After Cooled	
BORE AND STROKE	6.69" x 6.69" — 170 x 170 mm	
DISPLACEMENT	1413 Cu. In. — 23.15 Liter	
COMPRESSION RATIO	13.1 to 1	

DIMENSIONS AND DATA - 300 MODULE & 400 MODULE

WHEEL BASE	155"
LENGTH	291"
WIDTH	173" (24.5 x 32 Duals) 188" (30.5 x 32 Duals)
HEIGHT	146" (top of cab)
TURNING RADIUS	224" (center line of tractor)
WEIGHT	SA6D125/24.5 x 32 - 37,500 - 41,300 SA6D140/30.5 x 32 - 39,800 - 43,600 dry full fuel 60/40 Front/Rear ratio Maximum warranted weight 44,000
FUEL	550 U.S. Gals. Distributed equally in both sides of front frame.
ARTICULATION	40º Degrees
OSCILLATION	15º Degrees

CAB

Electric/Hydraulic tilt. ROPS and FOBS approved. Rubber mounted for sound and vibration dampening.

Precleaned and filtered air used for pressurization, eliminating dust and stale air.

Heated and air conditioned.

A fully adjustable deluxe seat is mounted on a Big Bud designed heavy-duty, locking swivel air-ride base.

AM/FM cassette stereo, tilt-telescopic steering wheel column.

INSTRUMENTS AND GAUGES

Engine tachometer, ground speed (mechanical), transmission gear indicator, engine oil pressure, engine water temperature, engine oil temperature, pyrometer, transmission oil temperature, transmission oil pressure, front & rear differential temperature, volt meter, air pressure, hour meter, clock.

WARNING LIGHTS

Fuel level, low voltage, engine oil pressure, transmission oil pressure, charge pump hydraulic oil pressure, engine coolant temperature, park brake, transmission oil temperature, engine air filter, hydraulic oil temperature, transmission oil filter, high beam indicator, low air pressure, sentinal.

DIMENSIONS AND DATA — 700 MODULE

WHEEL BASE	165 inches
LENGTH	309 inches
WIDTH	208 inches
HEIGHT	156 inches
TURN RADIUS	241 inches
WEIGHT	75,000 lbs.
FUEL	700 U.S. Gallons
ARTICULATION	40º degrees
OSCILLATION	15º degrees
TIRES	35.5 x 32

SPEED

Ground Speed 2100 RPM

1 - 2.6 mph	7 - 8.4 mph
2 - 3.2 mph	8 - 10.0 mph
3 - 3.9 mph	9 - 12.4 mph
4 - 4.8 mph	10 - 15.2 mph
5 - 5.7 mph	11 - 18.2 mph
6 - 7.0 mph	12 - 22.4 mph

1st R — 3.4 mph 2nd R — 6.2 mph

AXLES

Clark D-70500 outboard planetary
Total reduction 19.133:1
LTD slip (optional)

BRAKES

Air/Hydraulic (front axle only)
Equipped with micro-lock

TRANSMISSION

Full power shift, 12 speeds forward and 2 reverse. Electric/hydraulic controlled. A manually operated wet master clutch is used for vehicle "inching" in 1st gear, forward or reverse.

HYDRAULIC SYSTEM

Closed center, pressure compensated open loop system. 4 remotes are standard. Lever type (ISO) couplers. Pump capacity 60 GPM at 21 RPM with 2500 PSI maximum operating pressure. 135 gallon reservoir in rear frame.

ELECTRICAL SYSTEM

Starter System — 24 volt neg. ground All other electrical — 12 volt neg. ground
Alternator — 130 amp Battery — 2 each 8D heavy duty

Big Bud tractor models

Numbers Built

Model	Series	Engine	Model	Engine hp	Number built
HN250	1	Cummins	NT855	250	59
HN320	1	Cummins	NTA855	320	11
HN350	1	Cummins	NTA855	350	25
HN360	1	-	-	-	-
KT450	1	Cummins	KT1150	450	15
KT525	1	Cummins	KTA1150	525	2
HN320	2	Cummins	NTA855	320	22
HN350	2	Cummins	NTA855	350	1
HN360	2	Cummins	NTA855	360	28
KT450	2	Cummins	KT1150	450	32
KT525	2	Cummins	KTA1150	525	26
16V 747	2	Detroit Diesel	16V92T	750	1
8V400	3	Detroit Diesel	8V92T	400	2
8V360	3	Detroit Diesel	8V92T	360	1
320/10	3	Cummins	NTA855	320	9
360/30	3	Cummins	NTA855	360	14
400/20	3	Detroit Diesel	8V92T	400	10
400/30	3	Detroit Diesel	8V92T	430	33
400/30	3	Cummins	KT1150	400	8
450/20	3	Cummins	KT1150	450	1
450/50	3	Cummins	KT1150	450	31
525/50	3	Cummins	KTA1150	525	150
525/84	3	Cummins	KTA1150	525	2
600/50	3	Cummins	KTA1150	600	2
650/35	3	Detroit Diesel	12V92T	665	2
650/50	3	Detroit Diesel	12V92T	650	6
650/84	3	Detroit Diesel	12V92T	650	1
370	4	Komatsu	SA6D-125-1	370	2
400	4	Caterpillar	3406B	402	5
440	4	Komatsu	SA6D-140-B	440	2
450	4	Cummins	NTA855C450	450	6
450	4	Detroit Diesel	8V92TA	450	1
500	4	Komatsu	SA6D-140-B	493	2
500	4	Deutz	BF12L513C	500	1
740	4	Komatsu	SA6D170A-1	740	2

Big Bud tractor model styles

Round fenders	48 inch cab	Series 1	112
Round fenders	60 inch cab slanted radiator sides	Series 2	109
Half fenders	Big Bud Built Cab Cruiser cab	Series 3	169
Half fenders	ROPS cab	Series 3	104
Slanted tank ends	new style	Series 4	21

Transmissions used in Big Bud tractors

Make and series number

Clark	8821	3
Detroit Diesel Allison	60661	2
Fuller	RT & RTO 12513	239
Fujitech	ST1-079	18
Twin Disc	TD92-2610 TD61-1154 TD61-2613	253

Data supplied by Ron Harmon from the Big Equipment Co.

Big Bud Today!

From its humble beginnings in 1969, created by a need for a reliable, high-horsepower tractor, the Big Bud name has grown to be a household name all around the world. Havre, MT native, Ron Harmon, entered the picture in 1974 after Hensler and Nelson had built 20 tractors. Under his guidance, the company went on to build over 500 more units.

The great transmission shortage of 1979 caused severe financial hardship for the company. Although production of Big Buds continued, management and ownership had been shifted and things were never the same. Full production of the last Big Bud was in 1990.

In the years to follow, Harmon felt that his loyal customers had been given the back seat and weren't getting the attention that he felt they should have. These were people that had enough faith in him to buy his product initially and people that he felt he should still be providing service to. With the purchase of a Big Bud, you didn't just get a tractor; you got a product that Harmon and his team put their stamp of reliability and service onto.

In order to make Big Bud and its customers a priority again, Harmon formed Big Equipment, LLC. When the Big Bud series was designed, it was built on the philosophy that it was infinitely rebuildable. The frames were pre-drilled so that motor mounts could be changed to accommodate different engines. The power train was easily removable and exchanged. Common parts were used that were easily acquired through any parts distributor. What might appear to be a worn out tractor could easily be refurbished and put back to use as prime horsepower on the farm.

While the rebuildable concept was always in Harmon's mind, the tractors are now reaching the age where they are in need of being refreshened, repowered or even made bigger to meet the demand of today's equipment. Big Equipment's main focus is to take care of the Big Bud customer base with parts, service or whatever need they might have to maintain their equipment.

While Big Equipment also carries several short-line products, their lot is full of high-horsepower tractors. Many of those are Big Buds waiting for partial or full refurbishments. Harmon is proud to say they have taken in a couple 30,000 and 40,000-hour tractors and brought them back to life as new models. In many cases, these tractors can be put back into service better than new due to the advancement in technology and parts.

The most recent rebuild at Big Equipment consisted of a 650/50 Big Bud. When new, this tractor was equipped with a 12V92 Detroit. Through the years, the engine was changed to a Cummins, but the needs of the owner's farming practice demanded even more power. In order to meet those needs, Big Equipment made the modifications necessary to accommodate the install of a 1,000 hp 3508 Caterpillar engine. A new hood was built to fit the wider engine and the tractor was rebadged a Big Bud 950. The 35-year-old tractor rolled out the door capable of doing more work than a modern tractor and if needed, can be worked on by the owners.

Harmon states that the frustration is widespread with customers who have purchased new equipment. Of the four or five manufacturers who are building 250+ horsepower tractors, they own their own parts distribution, making repairs difficult for the tractor owner. These customers are asking for standardized components that can be purchased locally and installs that they can do on their own, limiting downtime and expense.

Moving forward, Big Equipment has made it their mission to provide parts, service, and partial or total rebuilds for Big Bud owners. They have moved their operation from a manufacturing base to a rebuild and service base. While current customers continue to return to Big Equipment, new customers are coming forward in search of a solution to their horsepower needs. With an infinitely rebuildable tractor, owning a Big Bud could mean that it's the last new tractor you'll ever need to buy.

While the Series 4 tractors were the last new Big Bud models to be built, Big Equipment is looking to the future with the design of a new model. Could Series 5 be around the corner? Time will tell, but one thing is certain: Big Buds need Big Equipment.

BigEquipment.com

16V-747

BIG BUD

Acknowledgements

The inspiration for my third book came from Stephen Moate, Publisher of Japonica Press who first introduced me to these giants of the land more than 16 years ago. Amongst all the big tractors built, Big Buds are both his and my favourites.

Bid Bud tractors are mainly confined to a very small area of North America around the home town of Havre, Montana where they were built.

These big tractors can be found in several other States of America and Canada and in several countries around the globe, but not in the concentration found around Havre where most of my research and photography was undertaken.

My largest source of information and help has been in Montana; I would like to thank all the people involved with this mammoth project. Each person or company in their own way has been equally important in the compilation of 'The Big Bud Tractor Story' either with historical or technical information or by supplying images or operating these big tractors for me to photograph.

There are a handful of people I would like to say a special thank you to, and they are Ron Harmon and his team at Big Equipment LLC, Robert and Randy Williams, Keith Richardson, Leo and Bart Bitz, Dan Sinclair and all his colleagues at Meissner Tractors Inc, Charlie and Burnie Inman, Toy Farmer magazine and Gina Harvey.

Both Big Equipment LLC and Meissner Tractors Inc have been invaluable in supplying accurate and detailed records of Big Bud tractor history.

If I have missed any company or individual off the list of credits, I hope they will not be offended as it was not done intentionally. I have talked with and met dozens of people in my search for information and images over the ten-year period it took to compile the 'Big Bud Story'.

Everyone has been so helpful and I thank you all once again.

There are some images that are essential to a book of this nature, which I have used and occasionally the photographer cannot be found. I have tried to trace the whereabouts and in some cases the names of photographers whose images I have found and used and have been unsuccessful in finding them, I hope they feel honoured to see their work published on these pages without their prior permission.

In alphabetical order, thanks to

Companies:
Big Equipment Co. LLC,
Meissner Tractors Inc.

Museums and archives:
Cass County Historical Society, North Dakota.
Montana Department of Agriculture.
Montana State University, Northern Archives, Havre, the Al Lucke Collection.
North Dakota Institute for Regional Studies & University Archives, Fargo.
Wisconsin Historical Society

Publications:
Classic Tractor magazine
John Deere Tradition magazine
Toy Farmer magazine.

Books:
Ultimate Tractor Power volume 1 - Peter D. Simpson
Ultimate Tractor Power volume 2 - Peter D. Simpson

The World Wide Web - www.
The Internet and the World Wide Web www have been a great source of information especially the various Montana sites which I used to source the Montana history.

Individuals:
All the Big Bud tractor owners who have spent time with me showing their pride and joy, a Big Bud tractor.

Mrs Freda Bafus, Bart Bitz and family, Leo Bitz and family, Peter Comben, Dave Curtis, Martin Fast, Brandon Gasvoda, Dick and Doug Hamilton and family, Ron Harmon, Steve Harris and family, Jason Hassert, Charlie Inman and family, Jon Lundwall, CF Marley, Paul Niederegger and family, Lorrri Norberg and family, Dennis and Joan Parker, Bruce Pester and family, Duane Oliver, Keith Richardson, Bernhard Roes, Dan Sinclair and family, Steve Stark, Michael Sullivan, Dieter Theyssen, Robert and Randy Williams and families, Red Wood, Jay Worrell, Randy Zimerman,.